British Mineralogy: or coloured figures intended to elucidate the mineralogy of Great Britain. By J. Sowerby (with assistance). F.P.

James Sowerby

British Mineralogy: or coloured figures intended to elucidate the mineralogy of Great Britain. By J. Sowerby (with assistance). F.P.
Sowerby, James
British Library, Historical Print Editions
British Library
1804-17
5 vol. ; 8º.
444.e.13-17.

BRITISH MINERALOGY:

OR

COLOURED FIGURES

INTENDED TO ELUCIDATE

THE MINERALOGY

OF

Great Britain.

BY JAMES SOWERBY, F. L. S.

HONORARY MEMBER OF THE PHYSICAL SOCIETY OF GÖTTINGEN,

DESIGNER OF ENGLISH BOTANY, AUTHOR OF ENGLISH FUNGI, ETC.

(With Assistance.)

As for the Earth, out of it cometh Bread, and under it is turned up as it were Fire. The Stones of it are the Places of Sapphires; and it hath Dust of Gold. JOB xxviii. 5, 6.

VOL. II.

LONDON:

PRINTED BY

R. TAYLOR AND CO., 38, SHOE-LANE, FLEET-STREET;

And sold by the Author, J. SOWERBY, at No. 2, Mead Place, Lambeth;
and by WHITE, Fleet-street; SYMONDS, Pater-noster-row;
and all other Booksellers.

TAB. CI.

FERRUM nativum*.

Meteoric Iron.

Class 3. Metals. *Order* 1. Homogeneous.
Gen. 7. Iron. *Spec.* 1. Native Iron.

SPEC. CHAR. Malleable, and nearly uncombined.

SYN. Ferrum retractorium, granulis nitentibus, matrici virescenti immixtis, (*Ferrum virens* Linn.) cujus fragmenta, ab unius ad viginti usque librarum pondus, cortice nigro scoriaceo circumdata, ad Plann, prope Tabor, circuli Bechinensis Bohemiæ passim reperiuntur. *Lithoph. Born. pars* 1. 125.

Stones said to have fallen from the Clouds. *E. King's Remarks on, &c.* 21.

Certain Stony and Metalline Substances which have fallen from the Atmosphere. *Phil. Trans.* 1802. *part* 1. 174. 183.

To introduce a substance, however curious, as having fallen like a meteor from the skies, or as Phaëton from the heavens, might seem absurd in a work on British Mineralogy. But, whatever may be the extent of this term mineralogy, it is

* We have arranged this as a native Iron, which is its great characteristic ingredient. It must come near the Iron of Siberia, Bohemia, &c., and be fol-

pretty universally understood to include a knowledge of stones and metals; among the latter of which we place this production, and feel much gratified in adding so great a rarity to the British catalogue. But we ought, in charity, to wish such may still continue to be rare, as otherwise the consequences might be dreadful. It is particularly to be noted that the same substances are only found under similar circumstances. Many of these stones haves fallen abroad in different places, but only two are known to have fallen in Great Britain; the first in Yorkshire, part of which is here figured; and the other in Scotland.

They have been found to contain

1. Silex.
2. Iron in a malleable state.
3. Magnesia.
4. Martial Pyrites.
5. Nickel.

The *silica* is lightish grey, in some parts rather vitreous, with rectangular yellowish fragments:—see *the left hand figure.* It is in very numerous but minute particles, which require the aid of a magnifier to be distinguished.—The iron is grey, much dispersed in particles of different sizes, mostly very small, often in rows, and sometimes in veins.—The magnesia seems combined with the silica, and the nickel chiefly with the iron.—The pyrites are dispersed in particles among the whole, some enclosing malleable iron, and some looking, when magnified, like particles of quicksilver; others are more distinct, and tarnished like common pyrites. They emit a blue blaze if projected on red-hot charcoal, and are easily fusible, becoming magnetic. The coating appears to be fused together, is very thin, and somewhat less magnetic than the rest; in some parts entering and forming veins within the stone. The whole is in texture like a compact sandstone, and may be crumbled into little pieces by the nail. The fracture is irregularly [illegible] earthy. There are dispersed through the whole several spherules of a laminated texture, which were first observed by Mr. Howard.

The upper figure is a fragment showing the coat and the indentations common to most of these stones, also the little reticulated cracks sometimes filled up with the whiter parts of the stone.

The right hand middle figure shows the other side of the same fragment, with a vein of iron, somewhat oxidated, since being broken; also little knots of iron, of a metallic lustre, which are irregularly scattered among the more minute particles of the same, with pyrites in the mass of the whitish earthy substance, composed of silex and magnesia.

The left hand figure shows the vitreous substance found in some parts of the stone, highly magnified. Count Bournon has found the same in the Sienna one. It is to be scratched with the nail, else we should have compared it with the peridot of Bournon, or chrysolite of Werner, which is found in the Siberian iron. It is remarkable, that besides this substance I have some crystallized pyrites adhering to a piece of Siberian iron in my possession.

The lower right hand fragment is magnified. It shows the granular formation of the stone, with somewhat tarnished pyrites, and the particles of iron in circular rows.

The two bottom left hand figures represent the earthy shperules*.

The following account of the Yorkshire stone was communicated by Major Topham:

The man who, by some fortuitous circumstance, happens to possess any extraordinary curiosity, has a very troublesome companion. He is liable to have his time occupied in

* These agree partly with those found by the Count de Bournon in the

answering letters from any stranger that may choose to ask questions, his house searched, his grounds ransacked; and, if the circumstance be very singular, he has the additional pleasure of having every word he says disbelieved on the subject. It was my good fortune to tumble into this predicament by a stone falling near my house in the country: and though I have been called upon, both publicly and privately, for a thousand accounts, and have answered innumerable inquiries from the ingenious, and those who had no ingenuity whatever; yet, as curiosity is a stream that never ceases, I found that this said stream might flow on for ever. I made one resolution, therefore, once for all. I was resolved to consign the stone in question to some public museum, where public curiosity might be amply gratified, and to deliver with it the most accurate account I was able to take from living witnesses on the spot, as I was at that time engaged on business in London. Mr. Sowerby, the publisher of a very ingenious work on mineralogy, has now the stone in his possession; and I doubt not the delineation of it, which accompanies this account, is extremely accurate and faithful.

The stone, therefore, coming from where it may, or bringing with it, in the words of Hamlet,

> "Airs from heaven, or blasts from hell,"

will no longer "blush unseen," but be subject to be examined, disputed, and commented upon by every philosopher in the United Kingdom, who may choose to visit the *Museum* of *Mr. Sowerby*; where they may, if possible, settle those points which I could never settle, though many of my inquirers seemed to think it mighty easy—viz: What projectile force could throw a stone of 56 pounds in weight from any volcano upon earth to the spot near my house where the stone fell? Whether it might not come from some volcano in the Moon? an idea to which French *Sçavans* the ground might not have power to conglomerate, to form at once, and, as it were, to knead together that heterogeneous mass of sulphureous and mineral matter of which this stone, and all others that are supposed to have so fallen, seem to be composed?

Having premised thus much, lest there should be found a person who might suppose I had the smallest wish or inclination to impose upon the world a wonderful story, I shall proceed to state what circumstances attended the falling of the stone in question, which was witnessed by many people who could have no interest in fabricating a false account, and were far too simple to have done so. What is most singular is, that it should have been so well attested, because on the high wolds of Yorkshire thousands of stones might have fallen, and there might not have been even a solitary shepherd, or his more solitary dog, to have witnessed the occurrence.

It was on Sunday, about three o'clock, the thirteenth of December, in the year 1795, that the stone in question fell within two fields of my house. The weather was misty, and, at times, inclining to rain; and though there was some thunder and lightning at a distance, it was not till the falling of the stone that the explosion took place, which alarmed the surrounding country, and which created so distinctly the sensation that something very singular had happened.

When the stone fell, a shepherd of mine, who was returning from his sheep, was about 150 yards from the spot; George Sawden, a carpenter, was passing within 60 yards; and John Shipley, one of my farming servants, was so near the spot where it fell, that he was struck very forcibly by some of the mud and earth raised by the stone dashing into the earth, which it penetrated to the depth of twelve inches, and seven afterwards into the chalk rock,—making in all a depth of nineteen inches from the surface.

While the stone was passing through the air, which it

did in a north-east direction from the sea-coast—numbers of persons distinguished a body passing through the clouds, though not able to ascertain what it was: and two sons of the clergyman of Wold Newton (a village near me) saw it pass so distinctly by them, that they ran up immediately to my house, to know if any thing extraordinary had happened.

In the different villages over which the stone took its direction, various were the people who heard the noise of something passing through the air, accurately and distinctly, though they could not imagine what was the cause of it: and in many of the provincial newspapers these accounts were published at the time from different persons.

In fact, no circumstance of the kind had ever more concurrent testimonies: and the appearance of the stone itself, while it resembles in composition those which are supposed to have fallen in various other parts of the world, has no counterpart or resemblance in the natural stones of the country.

The stone in its fall excavated a place of the depth before mentioned, and of something more than a yard in diameter. It had fixed itself so strongly in the chalk rock, that it required some labour to dig it out.

On being brought home, it was weighed; and the exact weight, at that time, was 56 pounds; which has been diminished in a small degree at present, by different pieces being taken from it as presents to different *literati* of the country. Mr. King, the antiquarian, in his account of *Sky-fallen Stones*, has published an account of this, with many curious and learned remarks on those which have fallen at different periods.

All the three witnesses who saw it fall agree perfectly in their account of the manner of its fall, and that they saw a dark body passing through the air, and ultimately strike into the ground: and though, from their situation and characters in life, they could have no possible detailing a false account of this transaction, I felt so desirous of giving this matter every degree of authenticity, that, as a magistrate, I took their accounts upon oath, immediately on my return into the country. I saw no reason to doubt any of their evidence, after the most minute investigation of it.

While Mr. Sowerby delivers in the work he is editing a very accurate delineation of the stone itself; at his request, I have transmitted to him this account of the circumstances attendant on it, to accompany the publication. But I mean not to enter into any literary warfare with those sceptics, who think it much easier to doubt every word of this account than to believe such an event could take place. Hume held the same language on miracles of a more sacred nature. There is no shorter way of disposing of any thing than to deny or disbelieve it: but sometimes

"They who come to scoff, remain to pray."

To perpetuate the spot where the stone fell, I have erected a pillar, with a plantation around it. The pillar is built over the exact place which the stone excavated, and has this inscription on a tablet:

Here
On this Spot,
December 13th 1795, fell from the Atmosphere
An extraordinary Stone!
In Breadth 28 Inches,
In Length 30 Inches,
and
Whose Weight was 56 Pounds!
THIS COLUMN
In Memory of it was erected by
EDWARD TOPHAM
1799.

Through the kind interference of my friend, G. Laing, Esq., Mr. Craufurd was so good as to send a part of the stone which fell in Scotland to compare with the Yorkshire one. It appears to consist of similar substances, and has the same sort of coating, though the pyritaceous particles are perhaps somewhat less conspicuous.

This stone was seen to fall into a small drain of water, at Possil Quarry, by two men, two boys and a dog, April 5, 1804. Among these was the overseer of the quarry, who was talking to a man in a tree near the place. At the time of its fall a noise was heard, which continued about two minutes, beginning in the west, and passing by the south, round towards the east, with as much noise at first as if three or four cannon had been fired near the bridge, which conducts the canal of Clyde and Forth over the river Kelvin, a mile and half westward of the quarry; and afterwards a violent rushing whizzing noise ensued*.

The overseer, upon observing the atmosphere, was alarmed at seeing a misty commotion, and called to the man in the tree, desiring him to come down, saying, "I think there is some judgment coming upon us." The man was scarcely got down, when something fell into the drain, splashing the water about 20 feet round. The eldest boy observed the appearance of smoke, and something reddish moving rapidly through the air from the westward. The younger boy, at the instant before the stroke against the ground, was heard to call out, "*O such a reek!*" He says he saw the appearance of smoke near the place where the substance fell. The overseer, when he observed a hole in the drain where the substance appeared to them to have fallen, made his arm bare, and thrust it into the cavity, which was not yet filled up, the water being very shallow. It was nearly perpendicular,

* The dog ran home as if frightened. The noise was heard by many people at different places, within 20 miles, and 30 feet under ground in the

or rather inclining from the west downwards to the east, at the bottom, about 18 inches deep. The overseer felt something hard, but was not able to move it with his hand: he therefore caused it to be dug out; and it proved to be the same substance as that sent from other parts of the world, and said to have been observed under similar circumstances.

Sir Joseph Banks first observed the similarity of these substances to one another when he went to see the Yorkshire one exhibited in Piccadilly, and compared it with a fragment of what he had got from Benares; and he was so good as to indulge me with the loan of a very perfect one from *L'Aigle* to compare with the British one: it was nearly black all over; but a small fragment had been detached, probably to see how it agreed with the others; which it does in all respects with the Yorkshire and Possil stones, except that the outside is of a purer black, perhaps from its falling in more favourable circumstances. The Yorkshire one fell against a damp chalk rock, and was partly discoloured by it. The Scottish one seems a little more rusty, and lighter, in the outer coat, the inside somewhat oxidated in spots; the natural effect of its falling through water. I have been favoured by Robert Ferguson, Esq., a friend of Mr. Laing, with another bit from *L'Aigle*, which is greyer than any others, as it has more iron in it, a largish vein of which forms a sort of reticulation, with a somewhat circular elliptical appearance, like some kind of marble, or irregular meshes.

The stone which fell December 13, 1803, in Bavaria, on a cottage, is somewhat remarkable for the day and month, agreeing with those of the Yorkshire one; and it is said that the pyrites were of a *cubical* form.

It may be a satisfaction to many of our readers to subjoin some account of other stones said to have fallen from the clouds, formerly and in our times, in foreign countries, from Mr. King's ingenious account, especially as they detail particulars that may develop some circumstances that ought to be known, and show more of the nature of these substances than are to be understood from the two which have fallen in our own country.

"Tradition has handed down to us the fall of stones in antient times. The learned Grævius leads us to conclude that the image of Diana was a stone which fell from the clouds. He tells us, on unquestionable authorities, that many other images of Heathen deities were merely such.

"Herodianus says that the Phœnicians had no statue of the Sun but a great stone which they reported to have fallen from heaven.

"Clemens Alexandrinus concludes the worship of stones to have been the first and earliest idolatry in the world.

"Plutarch mentions a stone which formerly fell from the clouds; and the old writer from whom he took his account says: 'It hovered about for a long time; seemed to throw out splinters, which flew around like wandering stars, before they fell, and at last it came down to the earth a stone of extraordinary size.' Pliny tells us of its being preserved in his days, and that it was of a dark burnt colour. He mentions one also which fell at Abydos, and was worshipped at that place; and of another at Pontidæa.

"Livy (whose credulity has been censured for preserving traditions of an extraordinary kind, which have been proved in ages of more enlarged information to be founded in truth,) describes a fall of stones on Mount Alba, during the reign of *Tullus Hostilius,* about 650 years before the birth of Christ."

Mr. King very properly quotes the Royal Psalmist: "The Lord also thundered out of heaven, and the Highest gave his Thunder, Hailstones and Coals of Fire*."

* Psalm v. At the same page he speaks of the large hailstones observed by my friend P. Rashleigh, Esq. of Menabilly, giving figures of them between

And if we read from verse 7 to 15, we seem to have in the language of the Psalms a tolerable account of the manner of their falling. "At the brightness which was before him, thick Clouds passed, Hailstones and Coals of Fire."

There are many opinions concerning the origin of these stones, and the Abbé Stutz remarks (vide Mr. King's account), "There is a great step from the disbelief of tales, to the finding out the true cause of a phænomenon which appears wonderful to us. And probably," says he, "I should have committed the fault into which we are so naturally led, respecting things we cannot explain; and have rather denied the whole history, than have determined to believe anything so incredible, if various new writings on electricity and thunder had not fortunately, at that time, come into my hands, concerning remarkable experiments of reviving *metallic calces* by the electric spark. Lightning is an electrical stroke on a large scale. If then the reduction of iron can be obtained by the discharge of an electrical machine, why should not this be accomplished as well, and with much greater effect, by the very powerful discharge of the lightning from the clouds?"

Thus, some consider that these stones may have been formed or generated in the air, by a combination of the mineral substances which had risen as exhalations from the earth; and the learned Mr. King gives us a very ingenious detail how it might naturally happen—"That the ashes from volcanos, after being thrown to an immense height, may consolidate by help of the particles of iron and pyritical dust, take fire either spontaneously or by means of the electric fluid within them, producing many explosions, and, by a sudden crystallization and consolidation taking place, form stones of various sizes, which fall to the ground; but the clayey ashes not hardening so rapidly as the metallic particles crystallize, an opportunity was given for impressions to be made on the surfaces of some of the stones, by means of others."

These dents, I think, look more truly like pieces having burst from the mass, which always appears more like an irregular fragment of a rock than a conglomerated body that had gathered in the air, which most likely would have been spherical. The Yorkshire stone, which has many of these hollows, fell alone.

We can say but little about the probability of their falling through our atmosphere from any of the planets: their coming hot to the earth, with so little velocity and force, after falling from so immense a distance, and their angular form, make it more astonishing. Mr. Howard observes that the concordance of facts seems to render it most indisputable that certain stony and metallic substances have, at different periods, fallen to the earth. Whence they came, he thinks is involved in complete obscurity.

In the account of the explosion of the meteor near Benares, in the East-Indies, by John Lloyd Williams, Esq. and the falling of stones at the same time, we find a good history of the nature and manner of their fall, which happened on the 19th of December 1798, at eight o'clock in the evening; and we conclude the effect must have been very conspicuous. A meteor is said to have appeared in the western part of the hemisphere, and was but a short time visible. It was observed by several Europeans, as well as natives, in different parts of the country, who described it as a large ball of fire, accompanied by a loud rumbling noise, not unlike an ill-discharged platoon of musketry. The light came into a Mr. Davis's room, projecting the shadow of the frames of the window as in the brightest moonlight. A number of stones were said to have fallen from it near Krakhut, a village about 14 miles from Benares, many of which were picked up from the fields. They penetrated about six inches into the ground, and were spread about 100 yards from each other; one, which weighed about two pounds, had fallen through the top of a watchman's hut. At the time the meteor appeared the sky was perfectly serene, and not the least vestige of a cloud had been seen since the 11th of the month, nor were any observed for many days after. These stones accord with those described by Mr. Howard, who observes there are no volcanoes on the continent of India.

If we suppose that these stones originate in the sphere of our globe, it must follow that the substances are within our atmosphere; and no doubt they are, or we should not have been acquainted with them. It has been thought that they may arise with vapours, smoke, &c., and by the attractive power of electricity become conglomerated; and that the inflammable part may have undergone combustion in a high region; and that, as it cools, the gravity being augmented, they are no longer driven by the currents which sometimes reign in the atmosphere, but, losing part of their velocity, drop to the earth again *. It may not be amiss to consider whether any other means may not be as natural, and this by degrees may perhaps lead to the truth.

That our travelling geologists have not found any thing concordant with this substance may be, because it would not be sufficiently remarkable to claim their attention; and so it may seem, when the appearance of this substance before ignition would not be new or uncommon. The rocks, of which I have got fragments, that I think most likely to produce such a combination of substances, are found in Wales; and as Scotland, and other places, even abroad, much resemble Wales in some particulars, they may do so in this; and, as they contain most of the substances which these do, and in the proper proportions, combined with other (perhaps

* Their force when falling is not very considerable.

fortuitous) circumstances (which happily do not very commonly combine to overwhelm us in a shower of stones), may be detached by the electric fluid attracted by the iron, which is known to be pure among pyrites. Thus, if we consider a part of the atmosphere surcharged with electric fluid coming in contact with such a rock, a discharge would take place; the iron would be heated to a certain depth; the opposition of damp, within the rock, would produce a rarefaction of sufficient violence * to cause a great report, and detach a fragment or fragments ; and the iron becoming so suddenly heated must set the sulphur on fire; which, while the stone was projected by the violent force necessary for detaching it, would be vivified as if blown by the bellows of a forge; till, the sulphur being exhausted, and the iron cooled, it would fall to the ground, with the scent of sulphur remaining, and sometimes a portion of heat. As it passed, a rushing noise (like the wind of forge-bellows) would be heard; the stone would emit sparks; and the irregularities would, more or less, cause pieces to fly from it, with a crackling or gun-shot noise; and it might sometimes have a comet-like form. The parts projected foremost would be of a white heat, as the force would augment it; and the sulphur would cause flame and smoke in clouds; the sparks and detached pieces would be left behind, as it passed, and appear something like the fiery tail of a comet. Thus far for conjecture: time may bring the truth to light; and if our travelling geologists will attend to this, I see no great improbability attached to the chance of finding rocks holding these substances in due proportion; and perhaps experiments may verify the truth of this or some other conjecture, and any natural *rationale* will be more or less confided in, as it agrees with the circumstances proved. Pyrites, we know, are always in action, or ready to act, under

* The force of steam is sufficiently known.

an infinite variety of circumstances. Thus, changes of weather, even without thunder or lightning, or atmospheric electricity, may produce, by uniting occurrences, similar phænomena.

Dr. E. Clarke, of Jesus College, Cambridge, entertains a different opinion respecting the origin of those substances; and is now employed in preparing a dissertation upon this subject for the Royal Society. The result of this opinion he has communicated, referring to his Memorial for the proofs necessary to establish its truth; as the work at large is too long for insertion here. He considers all the substances of the mineral kingdom as capable of existing in a solid, fluid, or aëriform state; according to the predominance of the active or passive principle: that is to say, of the principle of repulsion, or the principle of attraction. These two powers always counteracting each other, have been variously denominated; but their most recent appellations have been *caloric*, or the fluid matter of heat, and *the law of gravity*. The last of these was completely developed by Sir Isaac Newton; who determined the agency of the passive principle, or the law of gravity, to vary inversely as the square of the distance from the centre. The first, and perhaps the most important principle, whose agency prevents the particles of the most solid bodies from coming in contact with each other, remains to be developed by the discoveries of future science. One fact respecting it is generally admitted; that the power of attraction, independent of its agency, would be infinite. Admitting this truth, we can form no idea of the degree of solidity to which matter so circumstanced might be liable.

In regions remote from the earth's surface, where light enters into the least combination with matter, and generates the lowest degree of that modification of it, to which the term caloric has been applied, the particles of bodies, resulting from the decomposition of aëriform fluids, will be brought very near each other: and the consequence must necessarily be a degree of solidity equal to that of any known substance of the mineral kingdom. The mass of iron which fell in Sclavonia of 68lb. in weight*, Dr. Clarke considers as the result of such an agency in the passive principle. If this were really true, the iron so formed would be entirely destitute of carbon; which is found to be the case. There is no body more solid than water: yet water is brought by a similar process from the atmosphere, and it becomes fluid at the temperature to which the earth's surface is exposed. Iron may result from the combination of many aëriform fluids. Like water, it is capable of combustion; and there is not more reason for supposing it a simple substance, than there was for believing water an elementary principle, before it was discovered to result from the synthesis of hydrogen and oxygen.

We now give the analyses from the ingenious account by E. Howard, Esq. in Phil. Trans. 1802, part i. page 168 and following:

Of the stone which fell in Portugal, by the Royal French Academicians,

Sulphur - - - - - -	8½
Iron - - - - - - -	36
Vitrifiable earth - - -	55½
	100

Stone of Ensisheim, by Mons. Barthold, gave in 100 gr,

Sulphur - - - - - -	2
Iron - - - - - - -	20
Magnesia - - - - - -	14
Alumina - - - - - -	17
Lime - - - - - - -	2
Silica - - - - - - -	42
	97

Stone from Benares, the outside coating of which was found by Mr. Howard to contain iron and nickel. The pyritaceous part in 16 grains contained,

Sulphur - - - - - -	2
Iron - - - - - - -	10½
Nickel - - - - - -	1
Earthy matter - - -	2
	15½

The globular particles in 100 gr. contained,

Silica - - - - - - -	50
Magnesia - - - - - -	15
Oxide of iron - - -	34
Oxide of nickel - - -	2½
	101½

The earthy cement in 100 gr. contained,

Silica - - - - - -	48
Magnesia - - - - -	18
Oxide of iron - - -	34
Oxide of nickel - - -	2½
	102½

150 gr. of the Sienna stone.

Earthy part.

Silica - - -	70
Magnesia - -	34
Oxide of iron	52
Oxide of nickel	3
	159

In the metallic part,

Iron - -	6
Nickel -	2

150 gr. of earthy part of the Yorkshire stone,

Silica - - -	75
Magnesia - -	37
Oxide of iron .	48
Oxide of nickel	2
	162*

34 gr. malleable part.

Oxide of iron	37½
Nickel - -	4

55 gr. Bohemian stone. Earthy part.

Silica - - -	25
Magnesia - -	9½
Oxide of iron	23½
Oxide of nickel	1½
	59½

14 gr. malleable part.

Oxide of iron	17½
Nickel - -	1½

The specific gravity of the Yorkshire stone is 3508. That of the others is from 3352 to 4281.

Extracts from the hand-bill, which was given away at the time the Yorkshire stone was exhibited in London, and which is now preserved by Lady Wilson. Part is extracted from the *Oracle* of February 9, 1796, in a letter to Jas. Boaden, Esq.

The exact weight of the stone which fell, was weighed on being dug up. It was by Merlin's balance 3 stone 13 pounds: when taken up it was warm, and smoked. A labourer saw it coming down at the distance of about 10 yards from the ground; and, as it fell, a number of explosions were heard by three men at short intervals, about as loud as a pistol. At Bridlington, and at different villages, sounds

* When there is an overplus, it is from the metallic parts absorbing oxygen from the acids in the progress of analyzation.

were heard in the air, which the inhabitants took to be the noise of guns at sea. When the labourer recovered from the extreme alarm into which the descent of such a stone had thrown him, his first description was, "that the clouds opened as it fell, and he thought heaven and earth were coming together."

The following is the account given by Mr. L. Wilson: "I hereby certify to the public, that while I was in Yorkshire, near Capt. Topham's grounds, I heard noises in the air like the report of a cannon at a distance, and at the same time I felt two distinct concussions of the earth, which shook the buildings and the church near the spot where I was at the time. I was very much surprised, not knowing from what such circumstances could arise: within a very short space of time afterwards, I was informed that a stone had fallen within 200 yards of me; and a servant, belonging to my uncle, Mr. Wm. Parke, who resides near to Capt. Topham, was one of the people who saw it fall."

"Charles Prestin, son of the Rev. Mr. Prestin, eleven years of age, being in the church-yard at play, on Sunday, Dec. 13, 1795, at half-past three in the afternoon, after hearing a noise as of firing of cannon, heard at the above time a hissing in the air, and was sure something fell near the cottage belonging to Capt. Topham.

Given under my hand, this 29th day of April, 1796,

WM. PRESTIN,

Curate of Wold Newton, Yorkshire."

[illegible] Sowerby London

TAB. CII.

SILEX Quartzum, crystallizatum.

Crystallized Quartz; Cairn Gorum Crystals.

Class 2. Earths. *Order* 1. Homogeneous.
Gen. 4. Silex. *Spec.* 1. Quartzum.

SYN. Quartz hyalin plagiédre. *Haüy, v.* 2. 413.
Quartz hyalin rhombifére. *Haüy, v.* 2. 413.

CAIRN Gorum Crystals have been known for some years, and are said to have first caused the lapidaries to settle in Aberdeen, where they have been constantly employed in cutting them for seals, ring-stones, &c. They are perhaps of the oldest formation, and are found of various degrees of transparency, and sometimes are coloured yellow or brown. When of a deep colour they are esteemed as topazes *, and if clear and large are sold at a high price. The brown ones are also valued if clear; but when of a bad yellow, or muddy brown, the lapidaries have recourse to their art, and prove them to be rock crystals, by dissipating their colour, and giving them a transparent lustre. See *p.* 88.

This specimen is remarkable for the face *s* of Haüy (*i. e.* the little narrow face in the middle of the right-hand outline, which is often more regularly rhomboidal), being a truncation of the solid angle of the base of the pyramid; and the oblique face on the column, which is just below it,

* Topazes are found in the Brazils, &c.

on the same outline, corresponding with the faces on *the upper figure*, so as to make them more distinct. This latter is on the right hand of the column in most of the crystals on this group, and has not been before noticed. The next outline of a whole crystal lying on its column (from Cairn Gorum) has this face on the column on the other side, and a face on the edge of the pyramid and column, *m* of Haüy; which is somewhat rare. *The left-hand outline* has a little hollow in one corner, enclosing some liquid. *The next figure* shows the water, as it is commonly called, slightly magnified; which is mostly known by a little bladder of air moving as the crystal is moved. There is something that floats in the liquid, and looks like soot, or oxide of carbon. Crystals containing water or some liquid are sold at a high price. I do not know that any substance has been observed floating in the liquid within any crystal before.

The above group is in the collection of G. Laing, Esq. of Edinburgh. The others are in my own cabinet.

TAB. CIII.

QUARTZUM calcareum: *Var.* decomponens.

Decomposing calcareous Pudding-stone.

Class 2. Earths. *Order* 3. Aggregate.
Gen. 1. Quartzum. *Spec.* Calcareum.

Gen. Char. Quartz aggregated by the help of some cement.

Spec. Char. By the help of carbonate of lime.

It often happens that siliceous aggregates decompose, especially when their cement is calcareous; which is the case with the present specimen. It is perhaps now first spoken of, and gives us reason to suppose that the iron or colouring substance is disengaged from the calcareous cement by some agent capable of penetrating the inmost recesses of the stones. This cement filled the cavities of the mass, leaving them hollow, or with the porous remains of the pebbles only filling a part of the old cavity. It has been asked whether these stones are not rather forming than decomposing. We answer, that the cavities are shaped as if each had been filled by a whole stone, and the whole mass seems to be falling to pieces. In these masses some of the stones are whole, and in part soft, and may be scratched by a knife, or even by the nail, while the other part is as hard as a common flint pebble. Some of the pebbles are formed almost wholly of Carbonate of Lime, some of Silex, others partly of Oxide of Iron. I have some of the second kind from a well dug in Richmond-park, found among clay at the depth of 365 feet, taken notice of by my friend Mr. J. Murray, then gardener to Mr. Addington. I first received specimens of this nature from Warwick by chance; and the appearance of it was new to me, and to all whom I consulted about it. Lady M. Thynne, who was going to Warwick, kindly proposed to send me any mineralogical subject found in the neighbourhood. I requested her Ladyship to pay some attention to this, and was soon favoured with many curious specimens of decomposing rock; among which was that here figured. It is a piece of rock, chiefly quartz and carbonate of lime, inclining to be somewhat spongy and reddish with the oxide of iron: some of the cavities are empty, others have some remains of the pebbles, and others are nearly whole. *The lower figure* has the appearance of having been a common pebble, the colouring part of which has suffered oxygenization so as to become a loose ferruginous ochre, and the earthy parts of the stone are nearly separated. Some of the best mill-stones are of this nature on a larger scale, the quartz being left very porous; and these are preferred for grinding of corn.

TAB. CIV.

FERRUM sulphureum.

Sulphuret of Iron. Iron Pyrites.

Class 3. Metals. *Order* 1. Homogeneous.
Gen. 7. Iron. *Spec.* 4. Sulphuret of.
Div. 2. Imitative.

SYN. Sulphuret of Iron of particular Shapes. *Bab.* 204.

SULPHUR and Iron in combination are very common, and the forms of such compounds are extremely various, both in regular crystallization and in the ruder states. The specimen in *the upper figure* was given me by the Rev. Mr. Charles Sutton, and it resembles such as have been formed in chalk with an ochraceous outside. It has greatly the appearance of the fruit of the *Platanus Orientalis:* the surface, however, is crystallized, in quadrangular pyramids, or half relieved octaëdrons, with or without truncations. This may convince the discerning that the form does not depend on the fruit of the *platanus,* but is a form natural to itself, and curious for its regularity of appearance. The forms of the crystallization are magnified beneath, to show how curiously they connive, sometimes giving a floriform appearance. The next was found among the marly rocks at Sheppy Island, and seems to have been of a thickish consistence like treacle, dropping at intervals, and giving this singular formation of an inverted cone. The next figure, resembling a Mushroom-button, seems formed in a similar way, with trifling differences in the circumstances; as *the lower figure* does also, which resembles an Acorn. The crystallization and metallic appearance help to undeceive. However, this is not always the case; and even this might, from its rugosity, often having the appearance of the shagginess in a Mushroom, or of the squamæ on the cup of the Acorn, mislead those who do not pay regard to the crystallization natural to Pyrites. The cracks resembling stripes in *the left-hand Mushroom-like figure* help a little to favour the deception. It may not be amiss to observe, that on Sheppy Isle Mushrooms are extremely abundant; and this may have given rise to the idea of their being petrified, though any less fugacious Fungus might have stood a better chance of being petrified. This sort of Pyrites is extremely common in argillaceous marl all over the kingdom, and it varies infinitely.

TAB. CV.

FERRUM sulphureum, decomponens.
Sulphuret of Iron, decomposing.

Class 3. Metals. *Order* 1. Homogeneous.
Gen. 7. Iron. *Spec.* 4. Sulphuret of Iron.

SULPHURET of Iron, or Iron Pyrites, under certain changes of atmosphere, forms Sulphate of Iron, or Green Vitriol, as figured in *tab.* 23, and often falls to pieces with efflorescence; to prevent which, and to preserve curious specimens, we are obliged to keep them immersed in water. *The upper specimen* was sufficiently protected by the dark, partly crystallized, outward coat, from decomposing in the common atmosphere of a room, in a damp neighbourhood; but soon after it was broken, the more porous parts put forth curls of lightish-green vitriol, and at the edges chiefly, sulphur. It still slowly decomposes, and has continued to do so for some months: in time perhaps it may fall to pieces, or remain a porous iron ochre, like *the lower figure*. These specimens were found among a micaceous limestone in a quarry near Godstone, in Surry, where they roll down from the top in great abundance.

200

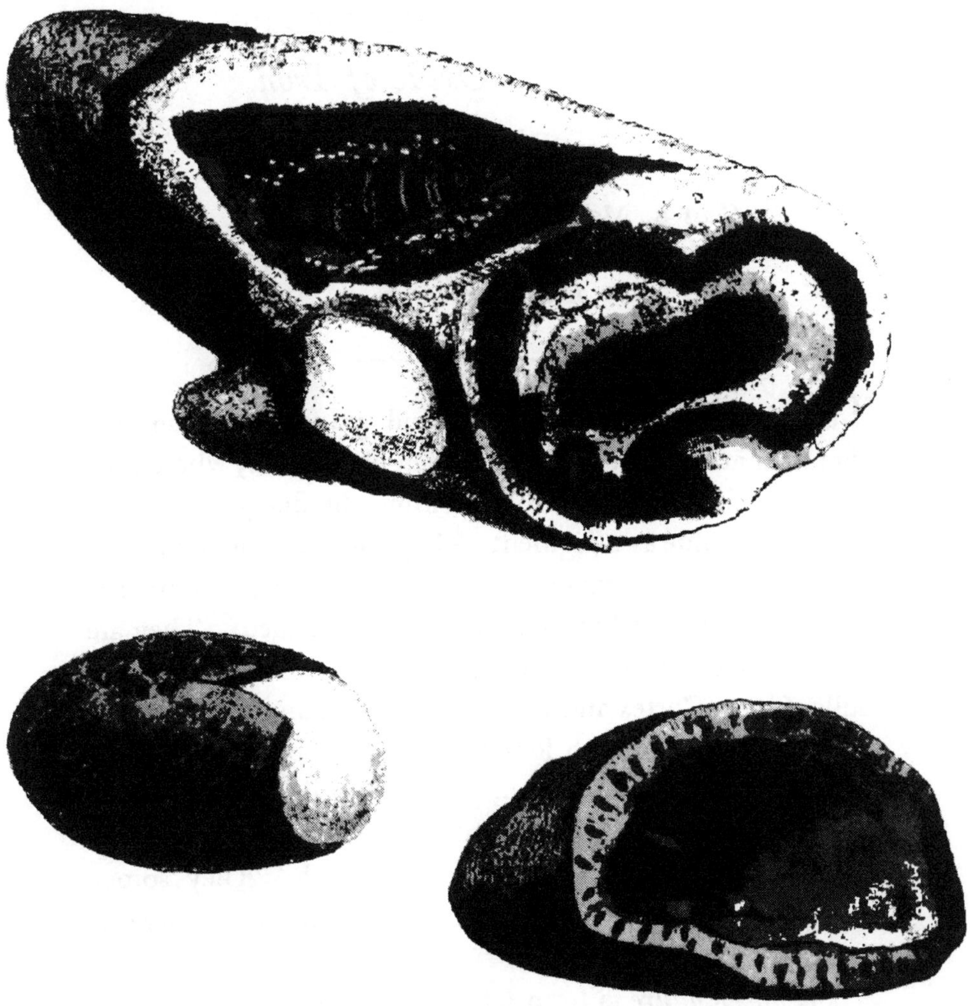

TAB. CVI.

FERRUM argillaceum.
Argillaceous Oxide of Iron.

Class 3. Metals. *Order* 2. Mixed.
Gen. 1. Iron. *Spec.* 1. Argillaceous.
Div. 1. Imitative.

THESE seem common in marly and gravelly land, and are abundant at Shotover hill and its neighbourhood, where they are situated so as to assist in forming the fine yellow ochre of so great value as a pigment. They vary extremely in their shape, sometimes branching like a stag's horn, or a branch of a tree, and have been taken for such petrified. They are often coated concentrically, imitating, as it were, the Medulla, Liber, Cortex and Cuticle. It may seem that the moisture in passing through loosish marl has been impregnated with the oxide of iron, and, periodically drying, leaves the marl and oxide of iron concentrated; which forms the coating, according to the looseness of the earth. They sometimes concentrate to a ball, but at other times have only one or two coatings.

The upper figure is from Charlton in Kent, and had the remains of a shell of the *Turbo* kind in it. The inside of the screw is covered with minute crystals of carbonate of lime: *tab.* 62. 63. There are other impressions of shells about it.

The left-hand figure has the form of a pebble with a lightish ferruginous ochre on the inside, and a dark crust.

The right-hand figure had loose pieces of ochre in it, of different colours, pinkish, &c., and they sometimes have wet marl and water in them. Such are called *Enhydros* by Sir J. Hill. The specimen came from Moushold Heath, near Norwich.

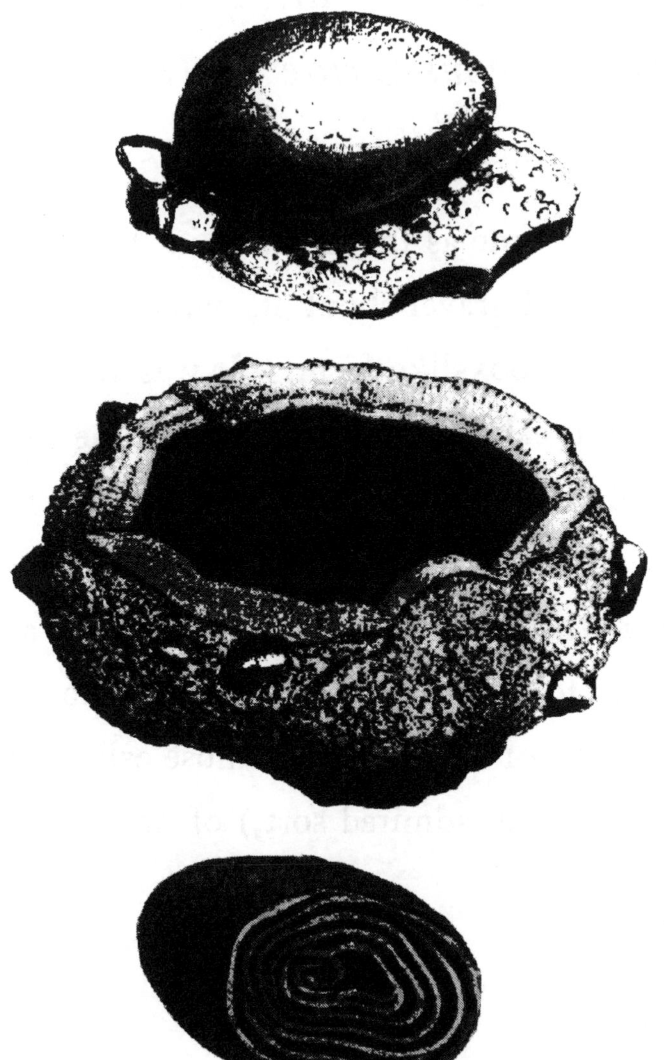

TAB. CVII.

FERRUM argillaceum.

Argillaceous Oxide of Iron.

THIS is nearly the same as that figured in the last plate; but it has pebbles and gravel about it, which are cemented with it into a hollow box-like form. It was discovered by an endeavour to break the stone from off the top of *the upper figure*, and thus formed a lid to the box. This was given me by T. Marsham, Esq. F. L. S. *The lower figure* is curiously concentrated with light ochre and a darker umber, and serves to give an idea of the formation of some pebbles, particularly of the nature of those called Egyptian pebbles, (not the most admired sort,) of which more will be said hereafter.

TAB. CVIII.

FERRUM argillaceum.

Argillaceous Oxide of Iron.

MY very good friend the Rev. J. Harriman sent me this from the banks of the river Tees, which place seems to abound in a great variety of curious petrifactions. It appears to be a fragment of a large flat piece of a confirmed Iron-stone, and, I should suspect, a very good one of the kind. It is however rendered additionally curious, as it includes Carbonate of Lime, or Calcareous Spar, in the form of Echinus spines, but different from any before noticed.

TAB. CIX.

BARYTES carbonata.

Crystallized Carbonate of Barytes.

Class 2. Earths. *Order* 1. Homogeneous.
Gen. 6. Barytes. *Spec.* 2. Carbonata.

I AM indebted to the munificence of F. Hall, Esq., for some superb specimens of Carbonate and Sulphate of Barytes. Among them was the present specimen of Carbonate of Barytes; and as it is the first time that it has been noticed with elongated spiculæ so distinctly seen in hexaëdral pyramids, I am pleased at the opportunity of exhibiting a figure of them. They are on a gangue of Galæna or Sulphuret of Lead, forming an irregular sphere, echinated as it were with these spiculæ: see *the upper figure.*

The lower figure on the right hand is a group of the spiculated crystals: on the *left hand* is a geometrical outline distinguishing the facets, three of which are continued to the apex, alternating with three which do not reach so high: sometimes these last are scarcely to be seen, and the spiculæ seem to have only three sides.

It is from Arkendale, near Richmond, Yorkshire.

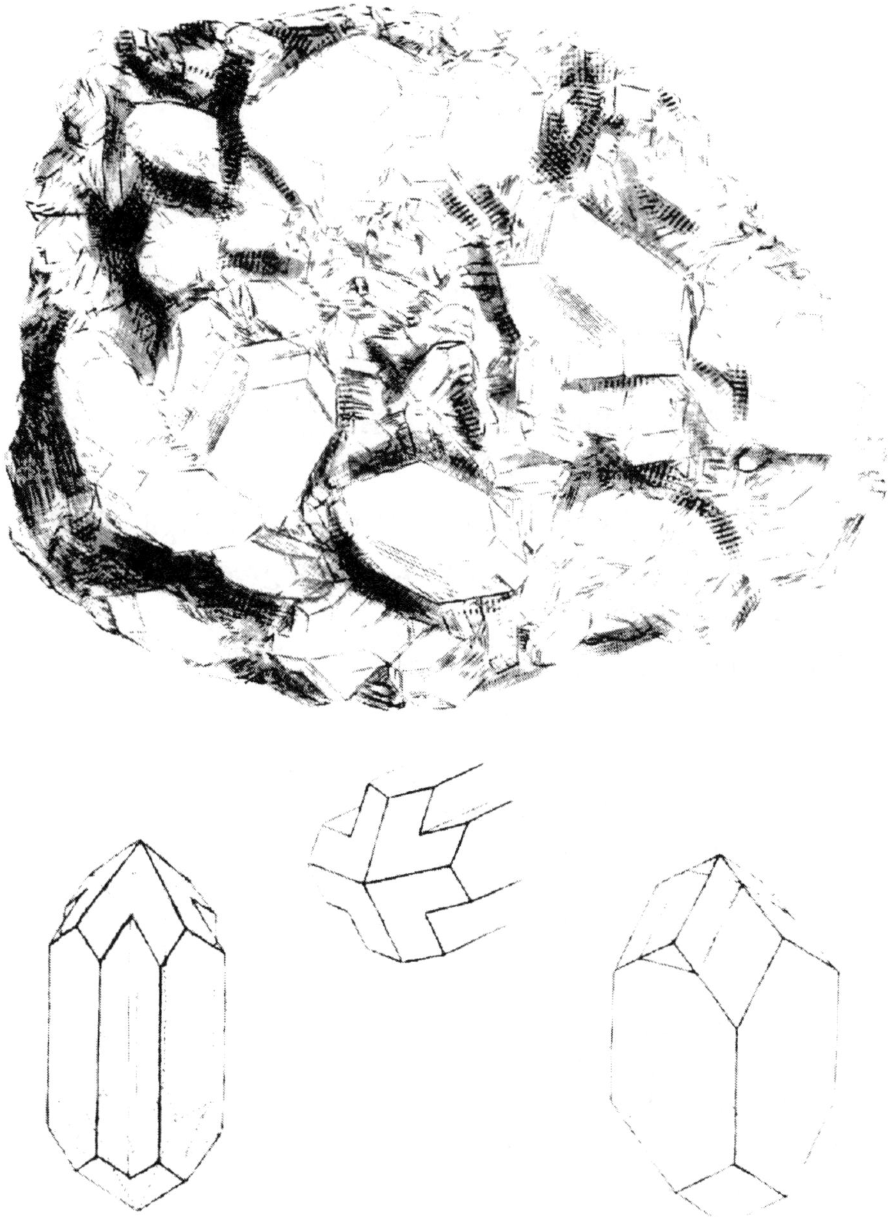

TAB. CX.

SILEX baryticus.

Harmotome or Staurolite.

Class 2. Earths. *Order* 1. Homogeneous.
Gen. 4. Silex. *Spec.* Barytic.

SPEC. CHAR. Combined with Barytes and Argil, fusible into a frothy enamel, with a greenish phosphorescence.

SYN. Staurolite. *Kirw.* 1. 282.
Harmotome. *Haüy*, 3. 191.
Kreuzstein. *Emmerling*, 1. 209.
Hyacinthe blanche cruciforme. *De Lisle*, 2. 299.
De Born, 1. 79.

BRITISH Staurolite has only been observed hitherto at Strontian in Scotland, a place famous for Carbonate of Strontian—see *tab.* 65. It is generally found on a gangue of Carbonate of Lime, which is mostly crystallized.

Its common form is a quadrangular prism, with the lateral solid angles truncated, and forming a four-sided pyramid, alternating at each end; or, in other words, making an elongated dodecaëdron, similar to that of garnet, but not of the same angle. Two of the opposite edges of the pyramids are mostly truncated, in British specimens. The crystals are generally larger than in those of Andreasberg, and more nearly resemble those of Oberstein.

Staurolite is chiefly admired for often assuming a cruciform appearance, looking like five crystals, four being united round a fifth. It appears however to be a regularity in the aggregating of the sides, without a sufficiency to fill up the lateral edges—see *the lower figure*.

They have been called twin crystals, as if two had passed across each other.

The whole appearance is somewhat glassy, of a blueish pearly lustre, having a foliated fracture on the broader faces. In other respects it is somewhat conchoidal, and hard enough to scratch glass.

Fusible by the blowpipe into a frothy enamel. It does not form a jelly when combined with acids. But if thrown on charcoal it emits a yellow phosphoric light.

On analysis by Klaproth it was found to contain

Silex - -	49
Baryt - -	18
Argil - -	16
Water -	15
	98

Its primitive form is said to be an octaëdron, divisible in the direction of the diagonals of the mutual base of the pyramids, so as to form four irregular tetraëdrons, or separate four solid angles, leaving a rhomboidal dodecaëdron, which might perhaps with more propriety have been called the primitive, to save confusion.

[illegible] Published by [illegible] London

TAB. CXI.

SILEX Quartzum; *Var.* opalinum.
Opal.

Class 2. Earths. *Order* 1. Homogeneous.
Gen. 4. Silex. *Spec.* 1. Quartz.
Div. 3. Amorphous.

SYN. Quartz resinite opalin. *Haüy*, 2. 434.
Opale. *De Lisle*, 2. 145. *De Born*, 1. 81.
Edler Opal. *Emmerl.* 1. 277.
Calcedoine irisée, Opale. *Daub.* 3.
Halb-opal. *Werner.*

WE cannot, I believe, boast of very beautiful Opals in England.

The rarity of this gem makes it worthy of notice, and the present specimen shows it passing from Calcedony to Cacholong, and from thence to Hydrophanous Opal, or what has been called *Oculus mundi*. Lastly, it forms the Semi-opal or Common Opal. The Calcedony is hardly to be distinguished in a drawing, it is the darker part mixed with Copper Pyrites in the figure. The name Cacholong applies to the opaque whiter and softer part, which may be scraped with the finger nail, and will, like all Agates and Calcedony, become more transparent in water. Hydrophanous Opal is the less opaque part, of a greenish hue, and in water becomes like the opal or blueish green part. In this state it resists the finger nail, and, while in water, is not to when dry.—Common Opal is nearly the same in appearance, either wet or dry, and comprises the grayish, blueish, greenish, and yellowish parts with a milky or waxy lustre, (like the Semi- or Halb-Opal of Werner,) with a vitrescent effulgence or yellowish fiery glare, in some lights, especially in the flaws. Fracture glassy. Hardness sufficient to cut glass.

The most beautiful specimen ever discovered of this substance is in the possession of J. M. Cripps, Esq. of Lewes, in Sussex. It was presented to that gentleman at Constantinople, by his excellency Count Ludolf, envoy from the court of Naples to the Porte. It belonged originally to the Capudan Pacha, and was found in Bulgaria. Unlike the common specimens, in which Opaline Wood appears in small veins intersecting the common Fossil Wood, or in a fragile state like Pitch Stone, it has throughout the whiteness of *Cacholong*, and in some parts the lustre and colour of the genuine Opal. It is larger than a man's body, and weighs 148 lb. 9¾ oz. avoirdupois. It presents half the trunk of a large tree, with the node of one of its principal branches. The timber, the bark, and every part of the mass is perfectly opaline.

The Right Honourable Sir Joseph Banks had a Mammoth's grinder in his possession, lately found on our coast, opalized. Other bones of the Mammoth are found in this country occasionally. I saw a tooth of a Mammoth from America, somewhat opalized, in the late Mr. John Hunter's museum. There are some also in the British Museum.

Sir Hans Sloane gave 200l. for an *Oculus mundi* now in the British Museum.

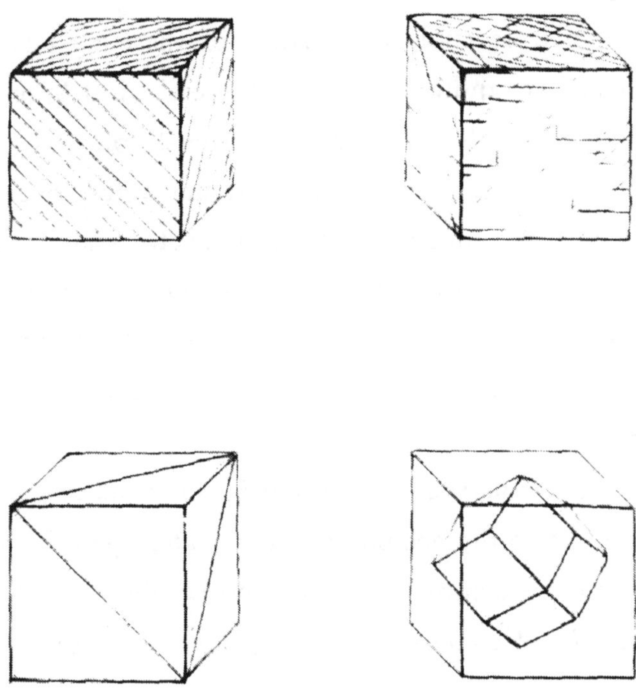

TAB. CXII.

ZINCUM sulphureum; *Var.* cubicum.
Cubical Sulphuret of Zinc, or Blend.

Class 3. Metals. *Order* 1. Homogeneous.
Gen. 6. Zinc. *Spec.* 2. Sulphuret of.
Div. 1. Crystallized.

THE rectangular or cubical formation of these crystals of Blend is, I believe, quite new to the Mineralogical World. Two specimens have been sent from Cornwall, one marked from Tincroft, the other from Polgooth, and were probably from the stores of a miner who did not exactly know from what mine they were brought. These luckily came into the hands of my kind friend A. MacLeay, Esq., Secretary to the Linnean Society, who was so good as to give me the one here figured.

The cubes of this specimen curiously show the diagonal striæ, and indicate the accumulation of plates upon the tetraëdrons, or more common modification*. *See these more plainly marked on the upper right hand geometrical figure.*

* It may not be amiss to remark that several other substances are striated in the direction of their principal modification; as *Fluor*, whose primitive is an octaëdron, but is generally striated in the direction of a cube; *Oxide of*

The right hand figure shows these marks more faintly like the original with the marks of the fracture parallel to the edges—see *the figure beneath*, which, if carried regularly on every edge, would produce the rhomboidal dodecaëdron, one of the characters of Blend. The gangue is mostly Copper Pyrites.

Since figuring this specimen I have met with a finer one from Cornwall with smaller distinct cubic crystals, or rather separate ones, much resembling Pyrites: the diagonal striæ however help to detect it, they are nearly gold-coloured, and iridescent externally, but dark like Blend within.

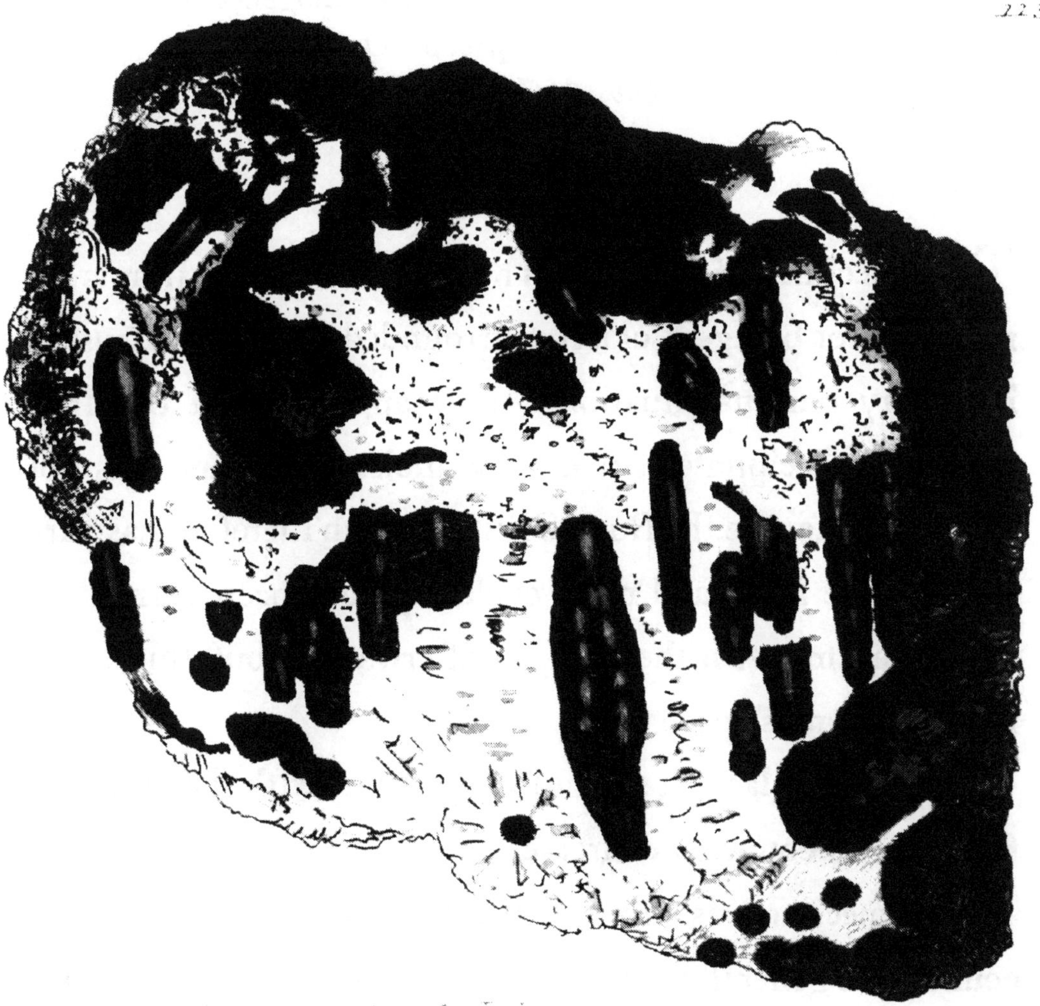

TAB. CXIII.

FERRUM oxygenizatum; *Var.* stalactiticum,
Stalactitical Oxide of Iron.

Class 3. Metals. *Order* 1. Homogeneous.
Gen. 8. Iron. *Spec.* 3. Oxide of.
Div. 2. Imitative.

THIS specimen of stalactitical hæmatitic Oxide of Iron is remarkable for being so closely surrounded with Rock Crystal, in a very curious situation. It is apparent that the Iron has been precipitated in a very moist state, and was not likely to penetrate the Rock Crystal, which therefore has surrounded the Iron, and must evidently have been formed at the same time with the Iron, or afterwards; but this latter is certainly most probable.

This is the blackest state of Hæmatites, with the radiating fracture, its common character.

This specimen is from Cornwall, the rocks of which country are generally supposed to be of primitive formation.

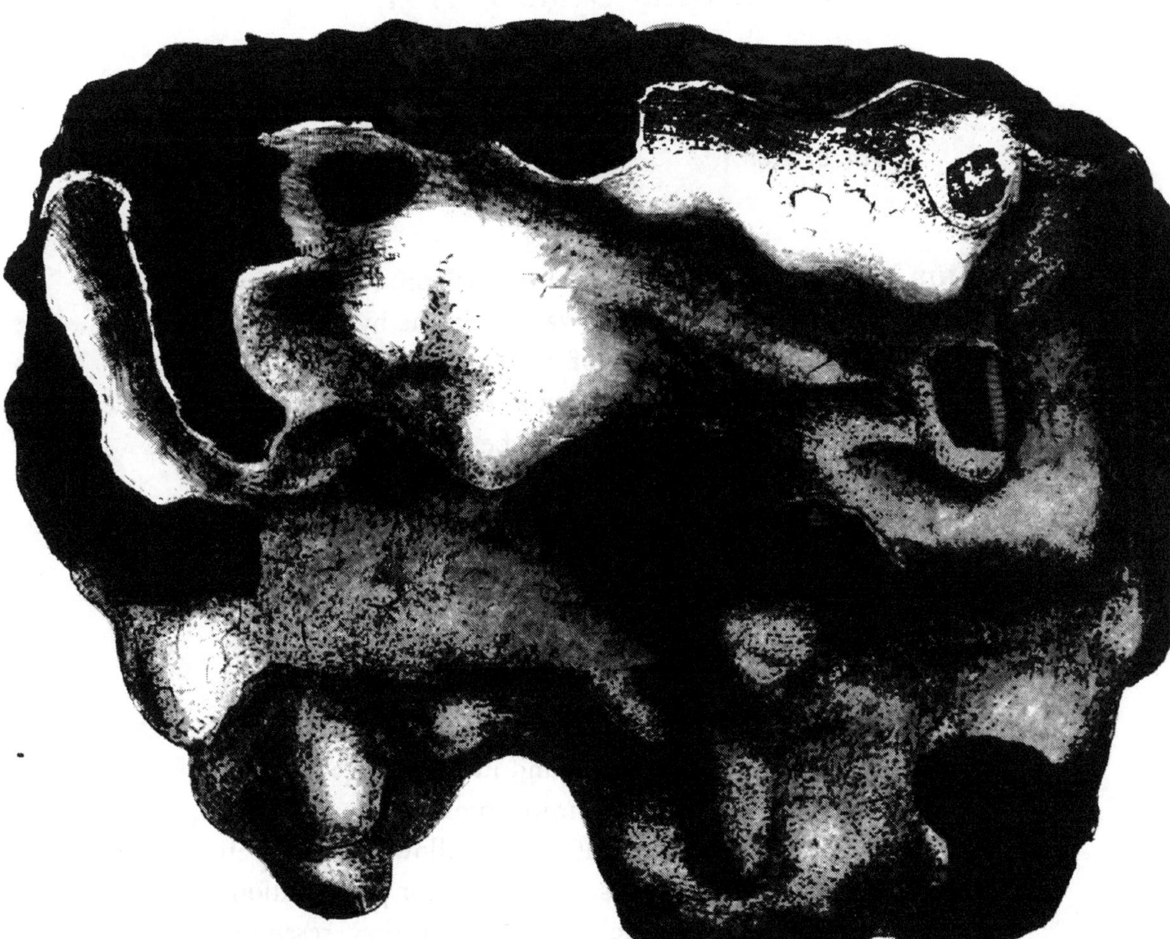

TAB. CXIV.

FERRUM oxygenizatum; *Var.* stalactiticum.

Stalactitical Oxide of Iron.

Class 3. Metals. *Order* 1. Homogeneous.
Gen. 8. Iron. *Spec.* 3. Oxide of.

This is another remarkable stalactitical production from Mam-Tor in Derbyshire. It was given me by my friend N. J. Winch, Esq., F. L. S. of Newcastle, and has been taken for Mineral Pitch, as its appearance would warrant us to suppose. It is however found to be an Oxide of Iron. Its situation is nearly as remarkable, when carefully examined, as that of the last; for its being so distinctly formed above the calcareous stalactite, gives an idea of its being a lighter substance. This might appear impossible: Nature, however, is seldom at variance with herself, and by a steady observance we may reconcile these apparent difficulties. It might happen that water very readily took up the Oxide of Iron which it found in its passage through the matrix, and, by dropping into some hollow, formed a stalactite, often coated by that which follows, and which was not so soon dissolved, or so soon fitted for precipitation. Some of its coats are in such a state that they resemble Martial Ethiops.

It comes near to the appearance of Hæmatites; it however has not the striated character, but rather the conchoidal fracture of Pitch. A little heat renders it magnetic, but does not take much of the blackness off: perhaps it contains a little Manganese, as the ochraceous part becomes somewhat crimson with heat.

A small degree of heat would melt it, if it were Pitch. This is a sort of proof of a natural application of a chemical trial necessary in Mineralogy.

The ochraceous part at first sight resembles a parcel of leaves from a chalybeate spring, and is of as deep an orange-colour as any I have seen.

The calcareous part may be the Calx sinter of Werner, which is a sort of crystallized stalactite, and the fracture is more or less distinct in this specimen, as well internally as externally. It seems to have been formed after the Iron, by nearly the same process, but was retained longer by the water.

TAB. CXV.

SILEX Quartzum.

Rock Crystal.

Class 2. Earths. *Order* 1. Homogeneous.
Gen. 4. Silex. *Spec.* 1. Quartzum.
Div. 1. Crystallized.

ROCK-CRYSTALS have been distinguished by many Mineralogists from the manner in which the crystals, depending on the column, swell or thicken in the middle. This mode of discrimination is perhaps tolerably correct, as I believe Rock Crystals alone are thus crystallized. It however does not determine that all Rock Crystals are thus formed, as may be seen by the Cairn Gorm Crystals, *tab.* 102. We may perhaps say that it can scarcely be otherwise determined from Quartz, unless by fire*; and it has been observed that Rock Crystals rubbed together are phosphorescent, or give light on collision in the dark, and exhale a peculiar empyreumatic odour. All stones as hard as common flint, to the consolidated aërial Diamond, will do this with very little difference.

Rock Crystal is often accompanied with Chlorite, and of so vivid a vegetable green that it much resembles moss, and has in some instances been taken for such: in this specimen, parts are impregnated with it, and seem decomposing: the whole has something besides of an opaque whitish cast, rather peculiar to this sort of Crystal. They often have the substance called Lac-Lunæ about them.

110

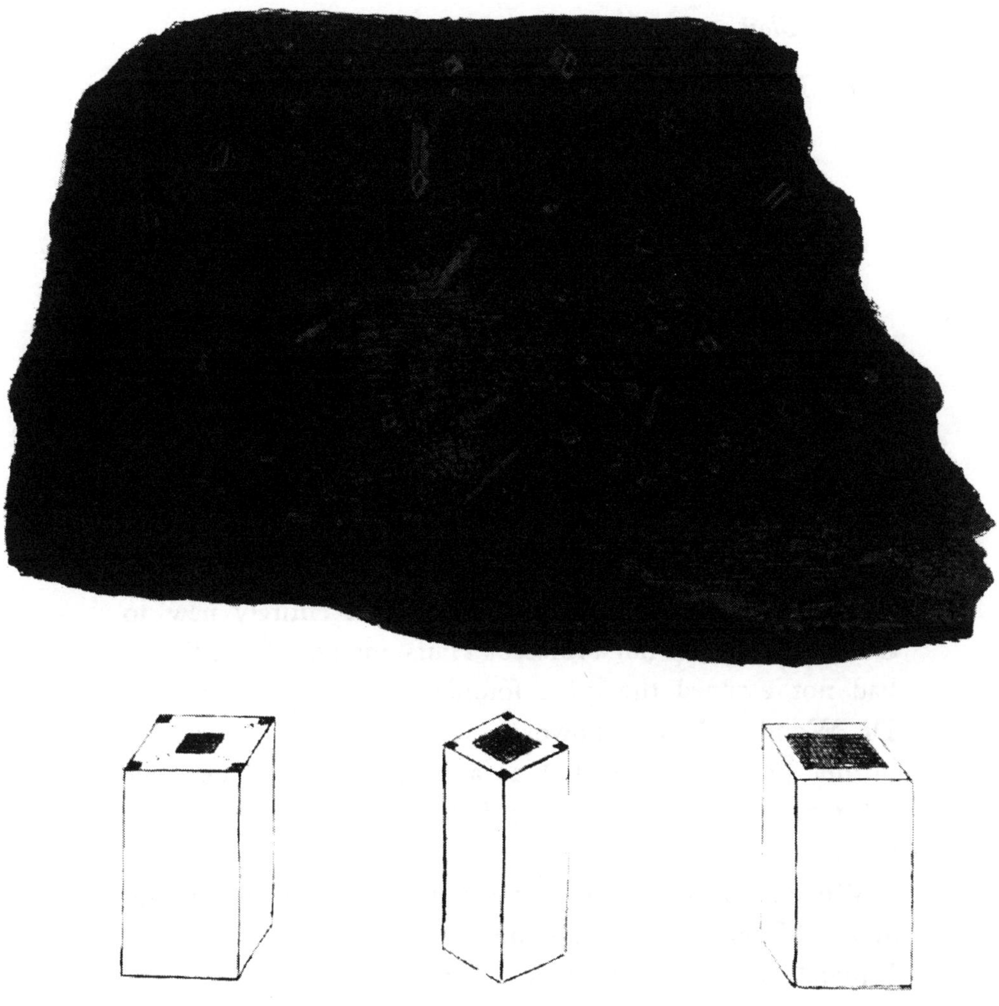

March 1 [illegible] Published by Jas Sowerby London

TAB. CXVI.

MACLE or *Chiastolite.*

Class 2. Earths. *Order* 1. Homogeneous.

Syn. Macle Basaltique, &c. *De l'Isle*, 2. 440.
Macles. *Daubenton*, 16.
Chiastolith. *Karsten*, 28.
Macle. *Haüy*, 3. 267.
A black slate holding in it great numbers of spinulæ of an Echinus spatagus*. *Woodward's Catalogue of Fossils*, *t.* 2. 55.

Mr. Davey having in the summer of 1804 found this substance on the summit of Mount Skiddaw in Cumberland, I am indebted to him for the power of adding it to the British list; and we had supposed it entirely new to Great Britain, had not Professor Hailstone assured us that it had not escaped the great founder of that professorship, Dr. Woodward, who, however, according to the knowledge of his time, considered it as a petrifaction. The same substance is said to have been found in some part of Scotland.

We do not know of its being mentioned by any author of the Mineralogy of Scotland, and indeed very few of the Mineralogists of the present day have mentioned it. The same sort, and in a similar gangue, has been brought to England from Bayreuth, and another sort has been imported occasionally from Lisbon, which are without a

* We have some Serpulæ petrified in marle, whose section much resembles

gangue, and much larger than these. They are religiously esteemed on account of the resemblance of the black part to a cross, and are sold by the monks as of great use to the possessor, generally somewhat mutilated to humour the appearance of the cross, and to fit them for wearing, as preservatives from all evils. Mr. Humphrys has a specimen of this last sort about three quarters of an inch broad at top, and nearly resembling ours.

Mr. Hailstone promised to lend me that in Dr. Woodward's collection at Cambridge, if different or better than the one figured; and as I have not received it, we may conclude it is nearly the same.

As far as we have seen, it is crystallized in slightly rhomboidal four-sided prisms of about 85° and 95°, according to De l'Isle. The laminæ seem parallel to the faces of the column, and there appears by the construction and meeting of the laminæ that there may be a diagonal division. Fracture fine, earthy, approaching to splintery. According to Haüy, it appears that the integrant molecule is the tetraëdron.

I hope we may meet with this fossil in greater perfection one day in Great Britain, so that I may be able to give a better account of it. I have something so nearly resembling it that I am apt to think they are the same, with an almost entire black centre, and merely a whiter pearly covering. The black is generally supposed to be like the shale or substance the crystals are enclosed in, so that the whiter part is chiefly spoken of; and it is very curious that they are so governed by a peculiar modification as to conjoin in such a manner as to envelope each other, yet retaining the rhomboidal form.

Spec. Grav. 2·9144.

As we have seen no analysis, we could not determine its systematic name or place; but we hope to attain more knowledge of it at some future period.

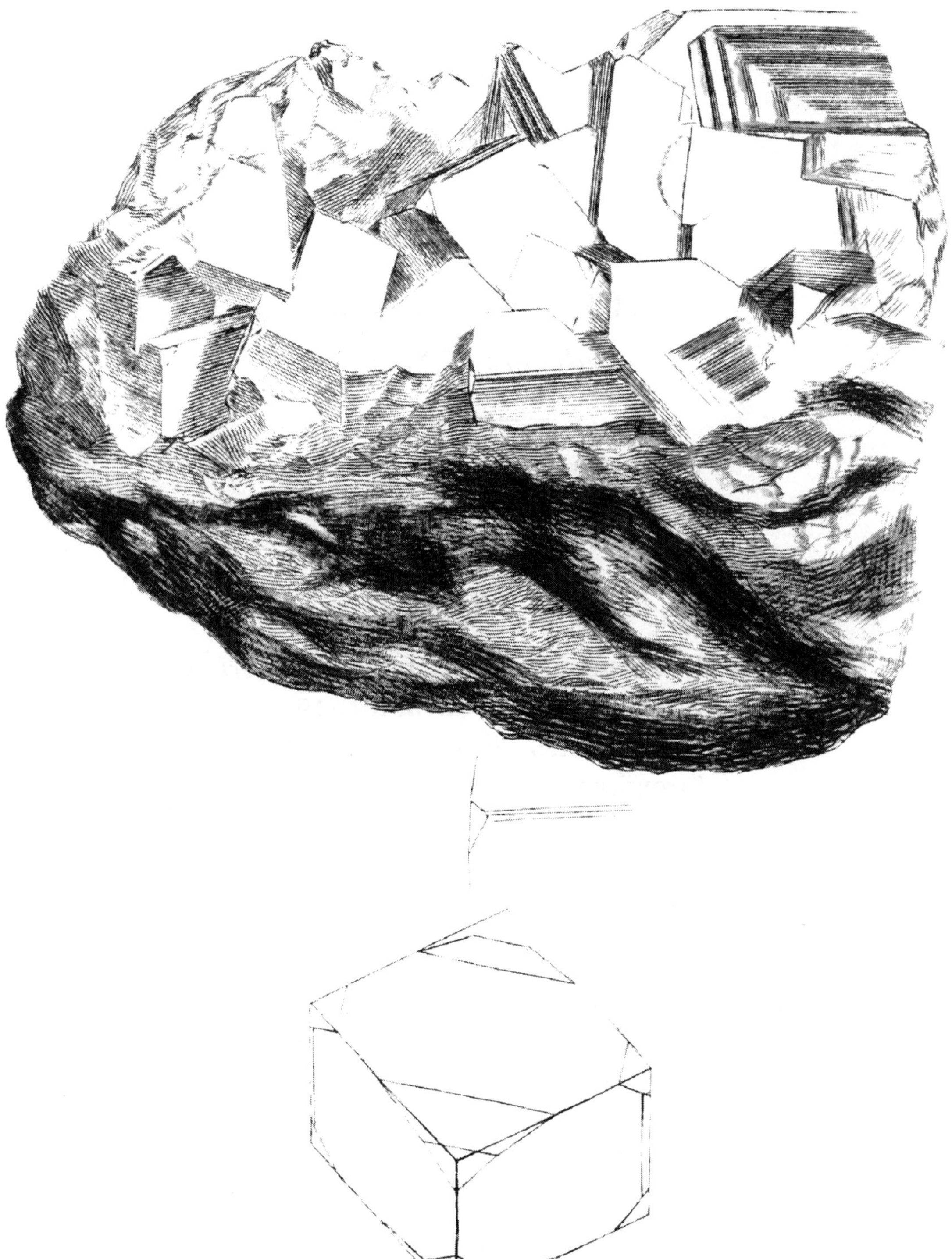

TAB. CXVII.

STRONTIA sulphata.

Sulphate of Strontian.

Class 2. Earths. *Order* 1. Homogeneous.
Gen. 6. Strontia. *Spec.* 2. Sulphate of.
Div. 1. Crystallized.

SPEC. CHAR. Strontian in combination with Sulphuric Acid.

SYN. Schewefel Saurer strontianit. *Emmerl.* 3. 312.
Strontiane. *Daubenton, tab.* 19.
La Cælestine. *Brochant,* 1. 640.
Strontiane sulfatée. *Haüy,* 2. 313.

SOME time in the year 1794 Mr. Tobin, of Bristol, first showed me, and was so good as to give me specimens of Sulphate of Strontian, from Redland, near Bristol; and it had been then but recently discovered, I believe, by himself, and was very little known. This was about the time Mr. Kirwan published his Elements of Mineralogy. We were next indebted to Mr. Clayfield for specimens, who, in 1797, discovered it at Aust Passage, near Bristol. The present figure is partly from one of his specimens. He observes that he discovered detached veins in different parts of the cliff. The strata in which the veins are found are nearly horizontal, consisting of Limestone of different hardnesses, and argillaceous Sandstone intermixed with Clay and Gypsum, and some of the fissures were filled up with

The primitive appears to be a rhomboidal prism of about 105°, and 75° according to Haüy. Ours are a little truncated at the solid angles, otherwise they represent the primitive, and are nearer so than any we have seen or heard of.

The fractures are parallel to the primitive faces, and those of the upper and lower faces are brightest, and often show a sort of opaline lustre. Sometimes we can discern the two diagonal fractures. Cross fracture somewhat splintery, harder than Carbonate of Lime, and softer than Sulphate of Barytes. The natural colour is a little milky and somewhat blueish, whence Werner called it Celestine. It is nearly transparent. Spec. Grav. according to Mr. Clayfield from 3·88 to 3·96; the fibrous about 3·91. Analysis by Mr. Clayfield:

Strontian -	116·5
Acid - -	83·5
A little Iron	
	200

Sulphate of Strontian was in many cabinets before this discovery of Mr. Tobin, as Sulphate of Lime, or Sulphate of Barytes,—among others in the Woodwardian Collection at Cambridge, as I am informed by Professor Hailstone.

It is found in Sicily, Mont-Martre near Paris, and in America. Those of Sicily are said to be the finest, and are columnar, not tabular like the Bristol ones.

116

TAB. CXVIII.

STRONTIA sulphata.

Sulphate of Strontian.

Div. 1. Crystallized.

THE plated Sulphate of Strontian here figured is in a sort of sandy matrix, and was the first specimen I received from Mr. Clayfield. It is like that which I first had from Mr. Tobin, excepting that the latter was in much larger pieces, without matrix, and nearly white. The crystals are mostly confused, yet are occasionally found very neatly determined at two or three sides, as in this specimen.

The geometrical sketch at the bottom shows the bevelling at the corner, between the primitive faces.

This kind is chiefly found at Redland near Bristol.

TAB. CXIX.

STRONTIA sulphata.

Sulphate of Strontian.

Div. 2. Imitative.

WE have represented the following varieties of this curious substance, to show those most generally known, at least in Great Britain.

The upper figure is from Aust Passage, near Bristol, and its curved appearance renders it at first sight sufficiently curious, besides its partaking of the blue and red colours.

The striated Strontian in *the middle figure* seems to answer to the name of Cœlestine, as Werner calls it, more generally than the last, as it is here coloured.

The fibres in this sort, which comes from Breslington, are very straight, and somewhat laminated.

TAB. CXX.

SILEX Granatus.
White Garnet.

Class 2. Earths. *Order* 1. Homogeneous.
Gen. 4. Silex. *Spec.* Granatus.
Div. 1. Crystallized.

THE Rev. John Harriman is the only person from whom I have received this substance, and it has not, to my knowledge, been spoken of by any author. Indeed its diminutive size might almost elude our search, even with the help of the lens. By that aid, however, we can discern in it the modification common to Garnet, viz. the rhomboidal dodecaëdron. We must own that the trial of the species was chiefly strengthened by the aid of the blowpipe, under which it resembles the common Garnet, *tab.* 43 and 44 of this work. These are found in irregular parcels, each crystal being from about the size of a small pin's head to extreme minuteness, sometimes pretty clear and bright, at other times yellowish and of a dirty hue. They have generally well defined sharp facets, which seem to vary but little. They run in lines at the intervals of the divisions in the matrix, which attracted notice by the curious appearance of its yellowish, greenish, light and dark reddish, and brown colours. These Garnets are sometimes mixed among a rough mass of nearly their own nature, which seems to incorporate with some Quartz. In fusion by the blowpipe they run into a black enamel without addition. The matrix is chiefly Carbonate of Lime, and a siliceous substance resembling dull reddish Jasper.

May 2 1806 Publish'd by Jas Sowerby London.

TAB. CXXI.

SILEX magnesiatus; *var.* amianthiformis.
Amianthus.

Class 2. Earths.	*Order* 1. Homogeneous.
Gen. 4. Silex.	*Spec.* Magnesiatus.

Div. 2. Imitative.

SPEC. CHAR. Silex in combination with Magnesia.
SYN. Amianth. *Kirw. v.* 1. 161. *Emmerl.* 1. 402. *Jameson*, 1. 442. *Wern.*
Asbestus maturus, Amianthus. *Waller*, 1. 408. 410.
Asbeste. *Haüy*, 3. 245.

AMIANTHUS is probably a decomposition, or change performed perhaps by some yet unknown chemical agent, as its situations in solid rocks seem to evince. This unknown chemical agent does not appear to be common, as the Amianthus is not an universal substance in the silky filamentous form. The present specimen was undoubtedly chosen with a scientific view by my kind patron Lord Seaforth, who brought it from Portsoy.

It is partly what is called Mountain Cork in a rather dense state, including some very fine transparent Carbonate of Lime, in which there are extremely slender filaments of Amianthus, seemingly passing through it, and in some parts so incorporated with it, that it is not to be discerned where they end or where the Carbonate of Lime begins. *The filaments* run perpendicularly to the more dense or cork-like surface of the upper and lower sides of the specimen, which however has more of a cottony texture, but is much finer than any *cotton* or *vegetable thread:* the first seems infinitely fine; the latter may be seen by a microscope to show a finally confirmed hollow filament*. It generally is found in Serpentine rocks.

* I was highly pleased to find in Mr. Carlisle's Croonian lecture read to the Royal Society, Nov. 8, 1804, that he had determined the final filament in muscular flesh. I had been much puzzled with the assertion that Lewenhoeck had found them infinitely divisible: having long since taken the pains to examine a piece of mutton flesh, I found the smallest filaments easily discernible, and according with Mr. Carlisle's accurate account, as far as I examined, but they are not so in *Asbestus*, or any filamentous earth.

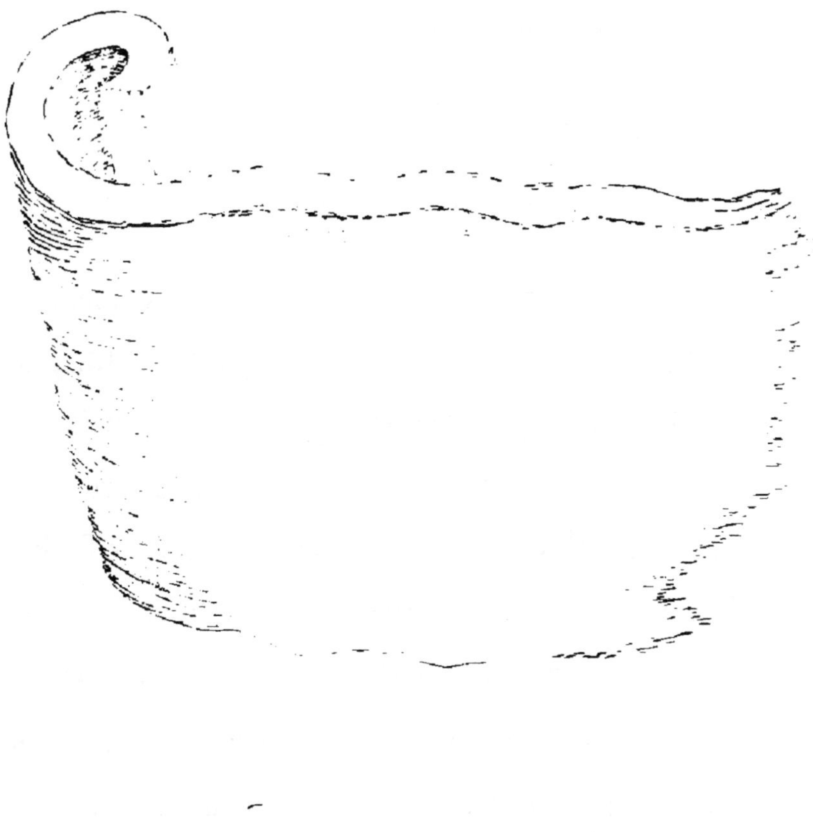

Nov 1 1806 Published by Ja.s Sowerby. London.

TAB. CXXII.

SILEX magnesiatus; *var.* amianthiformis.

Leather-like Amianthus.

Div. 2. Imitative.

THIS bears a strong resemblance to the Oak Leather or *Xylostroma Giganteum* of *English Fungi, tab.* 358, so as scarcely to be known by outward appearance, unless with the help of a microscope; when we find the one formed of exceeding fine filaments, and the other of hollow filaments. They are both flexible like leather, and to the touch equally soft. In cutting they require much the same force and sharpness of the instrument; they also tear nearly alike, with a shaggy filamentous appearance. One very sure test remains, that the one being heated red hot does not apparently undergo any change, but recovers its former appearance when taken from the fire; while the other burns away like touchwood or a fusee, or like the spunk prepared from *Bol. ignarius, E. F.* 132, or Fungus Tinder.

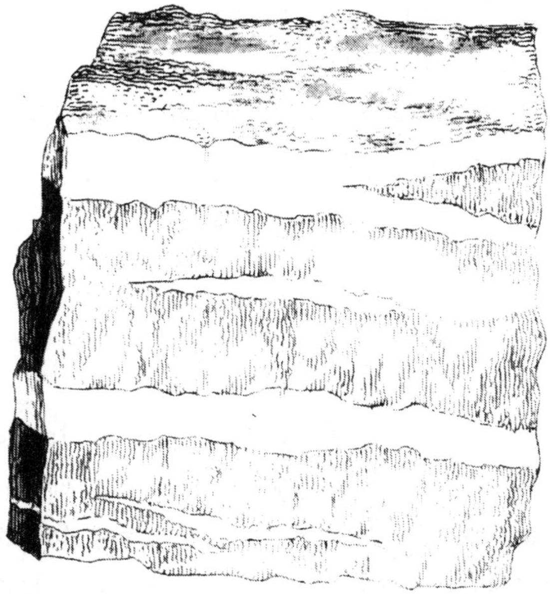

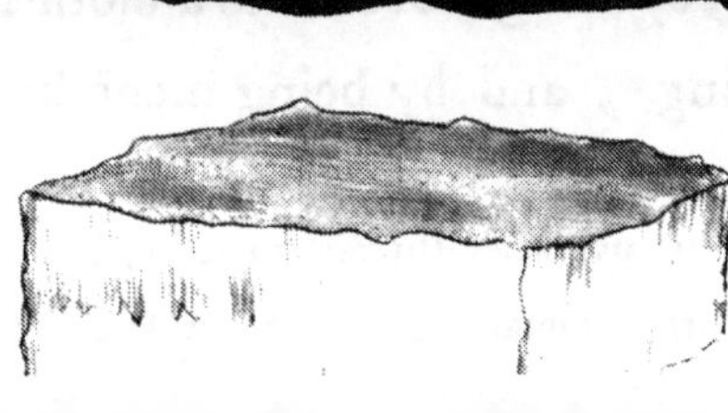

TAB. CXXIII.

SILEX magnesiatus; *var.* amianthiformis.

Silky Amianthus, or Asbestus.

Div. 2. Imitative.

THIS beautiful substance is found chiefly in the Island of Anglesea, North Wales; and at Portsoy, in Serpentine rocks, as they are mostly called. It is generally found in the fissures and cracks, passing like a sort of crystallization from the sides to the centre in infinitely small spiculæ, being sometimes quite indurated, though retaining nearly the same appearance as that which may be easily separated by the nail. *The upper figure* is from Portsoy, and is included in a somewhat woody Asbestos of a light colour. *The lower figure* came from Wales in a dirtyish Serpentine, and is partly covered with the Green Nephrite or Axe Stone, nearly approaching that from the Molucca Islands, of which the natives make their hatchets.

Amianthus, or Asbestos, as it has been more commonly called, was formerly used for preserving the ashes of deceased persons, by being woven into a cloth to wrap them up in while burning *, and by being incombustible it retained

* In weaving it they use other threads to assist; but those afterwards burn away, leaving a perfect amianthine cloth; a fine specimen of which was

those ashes most conveniently. Scotland and Wales have a satiny variety which runs in veins among Serpentines, and sometimes among a kind of Steatite. It varies in colour, but is most frequently white, satiny, and so much resembling silk that there can be no better comparison. It separates into silky filaments, of equal flexibility and fineness with the most attenuated thread, insomuch that they appear to divide beyond our power of examination. They seem solid, as do the filaments of all stones of this nature.

12.

TAB. CXXIV.

SILEX magnesiatus; *var.* amianthiformis.

Wood-like Amianthus, or Asbestus.

Div. 2. Imitative.

This is formed in upright and often curved filaments, sometimes in masses or plates. It is often indurated in a high degree, and resembles wood petrified—see *the lower figure*. The present kind is found at Portsoy in Scotland in great abundance and variety, crossing in many directions through the Serpentine rocks. *The upper specimen* was given me by Lord Seaforth. It is somewhat undulated, and varies in colour; is harder in some parts, and softer in others. In some it may be separated into flocculi with the finger nail, and in others it is so hard as to bear a good polish; when it is more brittle, and approaches to Actinolite. *The lower piece* was lent me by my friend G. Laing, Esq.; it comes also from Portsoy, and is remarkable for filling a narrow flaw in the Serpentine in a very thin plate, and showing fractures transverse to the striæ at nearly right angles. It is also very regular in hardness, so as to take a good polish all over the face; being about the hardness of common marble. This, and many earthy substances, resemble wood, especially the undulating slaty or irregularly schistiform rocks.

Publish'd by Ja.s Sowerby London

TAB. CXXV.

URANIUM oxygenizatum.

Oxide of Uranite.

Class 3. Metals. *Order* 1. Homogeneous.
Gen. 3. Uranium. *Spec.* 1. Oxygenizatum.
Div. 1. Crystallized.

GEN. CHAR. Dark brownish gray, dull, soft, brittle. Spec. Grav. 6·44. Difficultly fusible. Soluble in nitric acid.

SPEC. CHAR. Combined with a large portion of oxygen.

SYN. Uranitic ochre. *Kirw.* 2. 303.
Grun Uranerz. *Emmerl.* 2. 584.
Urane oxidé. *Haüy*, 4. 283.
Uran mica. *Jameson.*
Uranglimmer. *Werner.*

CORNWALL has produced this substance but very sparingly. It is mostly imported from Bohemia, Saska in the Bannat, and Saxony, on a gangue composed of Pech-blende* in a decomposing state, and was first discovered by Klaproth in 1789. It has been confounded by many with Muriate of Copper, and with Green Mica. Ours was labelled as Muriate of Copper from Cornwall, and it appears not to have been well known when Mr. Rashleigh published his first part of *Specimens of British Minerals*, &c., for it was there called "Thin four-sided Crystals of bright Green Copper Ore from Carharrack." Mr. Kirwan informs us of its being first taken for Green Mica by Werner, and afterwards for Calcolite.

The matrix of this specimen is Quartz, Arseniate of Copper, &c. Under the blowpipe it melts, becoming black.

TAB. CXXVI.

THIS variety is of a yellower hue, and appears to be splitting and decomposing, as it seems to decay at the edges, becoming rounded as if worn by oxygenizement. The gangue is chiefly a black Oxide of Copper.

Its Spec. Grav. is 3·1212.

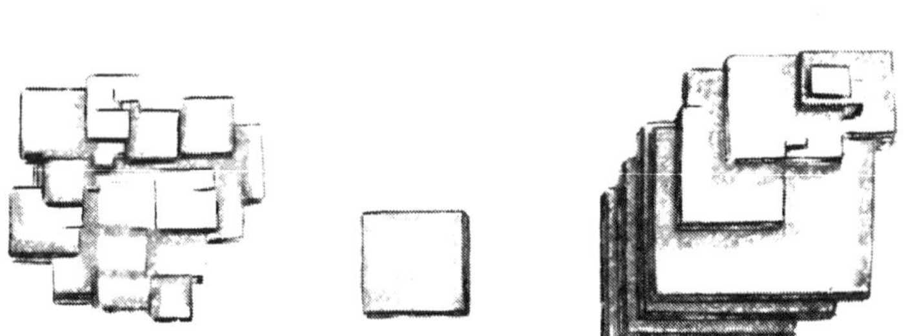

127

TAB. CXXVII.

BARYTES carbonata.

Carbonate of Barytes.

We are obliged to F. Hall, Esq., of Arkendale near Richmond, Yorkshire, for the superb specimen from part of which this figure was taken: it is at first sight very interesting, as the crystallization is apparently a most perfect representation of the dodecaëdral Quartz, with the two hexaëdral pyramids meeting at their mutual base without any intermediate prism—a very rare circumstance in Quartz. On further examination we find a modification seemingly peculiar to itself, and which has never been recognised in Quartz, or in any other substance before: see *the right and left hand figures*. It is so new in its nature that an expression is wanted for it, as it is not what may be termed mackled, but may rather be called articulated or jointed. These figures have one or two joints, lengthening the crystal endways without a tendency to continuing at the mutual bases, or filling up the interruption so as to form the plane of the column; the vacancy is just contrary to those of the Carbonate of Lead, which it somewhat resembles in the first instance: see *tab.* 89, 90, 91, and may be of much utility for external discrimination. We have been able by this specimen to determine partly the measure, and compare with Quartz—the incidence of one of the pyramids of Quartz upon the other measuring, according to Haüy, 103° and some seconds, and we find it in Car-

The Carbonate of Barytes has, in this specimen, a tendency to crystallize in groups, in a stellated manner: see *the middle figure*. The whole of the Carbonate of Barytes is chiefly formed on Carbonate of Lime, and is partly covered with Sulphate of Barytes in fine spiculæ. Some of the Carbonate of Lime is in Garnet dodecaëdrons with short columns, an uncommon form for Carbonate of Lime: see *tab.* 128, *right hand outline*.

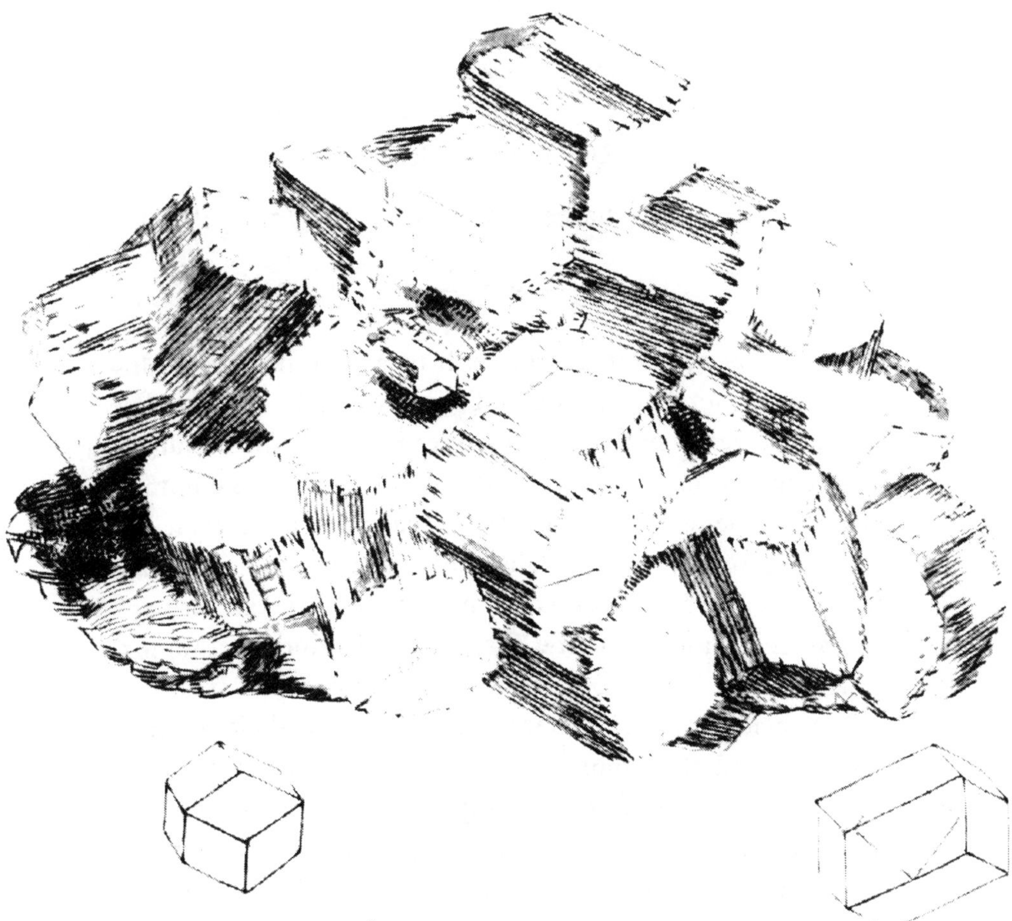

June 1 1805 Published by Jas Sowerby London

TAB. CXXVIII.

CALX carbonata; *var.* dodecaëdra.

Dodecaëdral crystallized Carbonate of Lime.

Class 2. Earths. *Order* 1. Homogeneous.
Gen. 1. Lime. *Spec.* 5. Carbonate of Lime.
Div. 1. Crystallized.

SYN. Chaux carbonatée bisunitaire [illegible]: *Haüy*, 2. 142.

SOME Carbonates of Lime, with regard to their crystallization, are not uncommon. This however is of the scarce kind; and was it not for the manner in which the modification appears to be formed, it might, under the common idea of its being a rhomboidal dodecaëdron, be esteemed as a Garnet, or some other substance that commonly exhibits that figure.—The fracture of Carbonate of Lime is very apparent in it; as well as the shapes of the primitive rhombs, which will be found to agree with the figure of the fracture; and are placed in such a manner as to form the dodecaëdron. Thus we find the column is formed of laminæ placed on the face of the rhomb decreasing from the lateral edges. Its termination is at one or both ends, clearly in the form of the equiaxe.—*The middle figure* expresses this with the fracture, which is mostly on the edges. The whole is a sort of inverse crystal; the column being, as it were, formed contrary to the usual custom.

The specimen from which this figure was taken was lent [illegible]

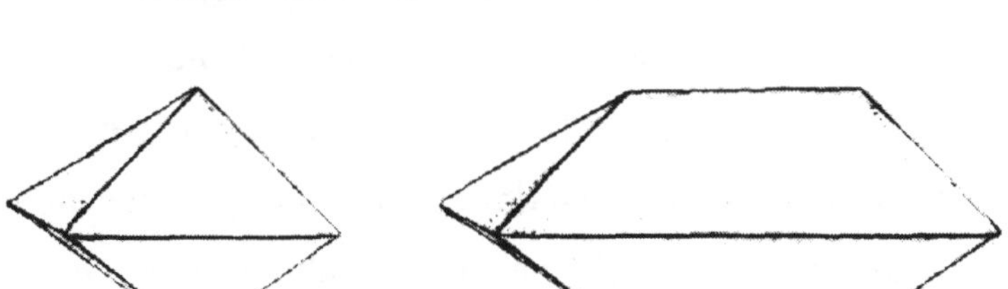

[illegible] Published by [illegible] London

TAB. CXXIX.

PLUMBUM sulphatum crystallizatum.
Crystallized Sulphate of Lead.

Class 3. Metals. *Order* 1. Homogeneous.
Gen. 15. Lead. *Spec.* 4. Sulphate of Lead.
Div. 1. Crystallized. *Var.* 1. Primitive.

SYN. Plomb sulfaté primitif P. *Haüy*, 3. 504.
Naturlischer blei vitriol. *Emmerl.* 3. 413.
Native Vitriol of Lead. *Kirwan*, 2. 211.

I BELIEVE there is no place in the world, except Anglesea in North Wales, where this substance has been found; and all we have heard or seen of it came from the Parys mine. We were extremely lucky, after having met with the best collection of crystals in the world, from the present octaëdron, through various modifications, to have found an extraordinary addition in another set that we had received through the favour of our good friend, and friend to Natural History, the Rev. Hugh Davies, F. L. S.*

These were so new as to make the former set doubly valuable. The present specimen is not so brilliant as many, but, of the kind, is one of the largest and best crystals yet procured. Its fracture in some directions is laminated, parallel to the primitive, in others vitreous. Perhaps one of its characters may be taken from its odour, which is similar to the smell of the place in which white lead is prepared, with a peculiar pungency. It is easily reduced on charcoal.—The forms are in general somewhat neat, and the crystals are sometimes as clear as the finest glass; at other times they are coloured with an oxide of Iron, from a yellowish hue to a rusty brown.—They are brittle, and soft enough to be scratched by Sulphate of Barytes, but not Carbonate of Lime, and are mostly found on an ochraceous gangue.

TAB. CXXX.

PLUMBUM sulphatum.

Crystallized Sulphate of Lead.

Div. 1. Crystallized.

THIS is one of the set of modifications of Sulphate of Lead which does not appear to have been noticed in any collection but my own, and shows a curious change as to the formation taking place by means of spiculæ. The inner part of the crystals are of the primitive octaëdron, apparently rounding to a new modification by decomposition, and forming spiculæ from the mutual bases of the two pyramids of the octaëdron.—This seems quite new to the mineralogical world: however, I have Carbonate of Lead in dodecaëdral crystals, looking as if decomposing, with spiculæ attached in a particular manner.

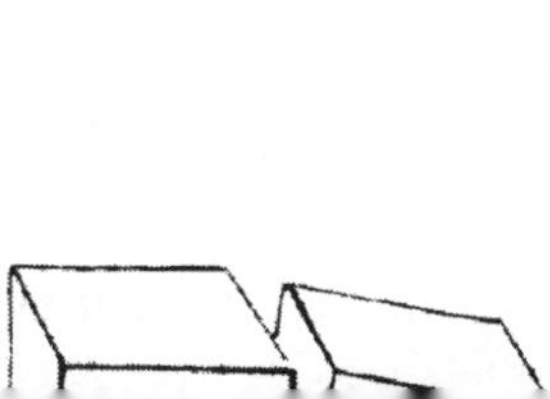

TAB. CXXXI.

PLUMBUM sulphureum; *var.* hæmatiticum.

Hæmatitic Sulphuret of Lead, or Hæmatitic Galena.

Class 3. Metals.	*Order* 1. Homogeneous.
Gen. 15. Lead.	*Spec.* 3. Sulphuret of Lead.

Div. 2. Imitative.

I HAVE heard it remarked from time to time that Galena was to be found in Coal at Matlock in Derbyshire, and I have had a specimen sent me from thence, but it was so full of a decomposing Pyrites that it fell to pieces.

Lady Aylesford, whose ingenuity is well known, and who has a fine collection of drawings of nearly all the English plants, made by her own hand, was so kind as to exert herself, when in the country last autumn, and procure for me some very curious minerals from Coventry; among which was the specimen here represented. This is not only rare from being among Coal, but for being in this curious hæmatitic form, a form which has not yet been noticed in Galena, and which will help more perfectly to show the nature of substances that have a regular primitive crystal, passing into, or placing themselves in, a rounding figure.

The present specimen is radiating from a centre, forming circular and hæmisphærical segments, much resembling Hæmatitic Iron Ore—*tab.* 56 and 131.

This substance is evidently formed of cubes, and has

and settling from its solvent with perfect preservation of its nuclei, which are very clear upon a careful examination. Although very minute, they are distinct cubes: and as any form made up of perfect cubes with level sides would, in forming a radiating figure, be loose in its texture, unless well closed with others in the interstices; so this, so loosely formed in that particular, is very useful to us to determine other facts of the like nature. *The left hand figure* is of two nuclei—and *the right hand figure* shows the manner of its accumulation, and will be useful in analogy.

Aug 1 1806. Published by Jas Sowerby London

TAB. CXXXII.

PLUMBUM sulphureum; *var.* antimoniatum.
Antimoniated Galena, or Sulphuret of Lead.

Class 3. Metals. *Order* 1. Homogeneous.
Gen. 15. Lead. *Spec.* 3. Sulphuret of Lead.
Div. 2. Imitative. *Var.* Antimoniated.

SYN. Plomb sulfuré antimonifère. *Haüy*, 3. 462.

THIS is often found accompanying common Galena. It is externally known to Mineralogists by the peculiar small striated appearance which is attributed to the nature of the crystallization of the Antimony, and which seems to derange the primitives of Common Galena; so that one substance interrupting another may be recognized here; and perhaps in some other place we may be able to show something more important in the nature of the laws of mixed crystallization.

The specimen here figured was sent by my friend the Rev. J. Harriman. It seems to be less antimoniated towards the base, and somewhat resembling *tab.* 131.

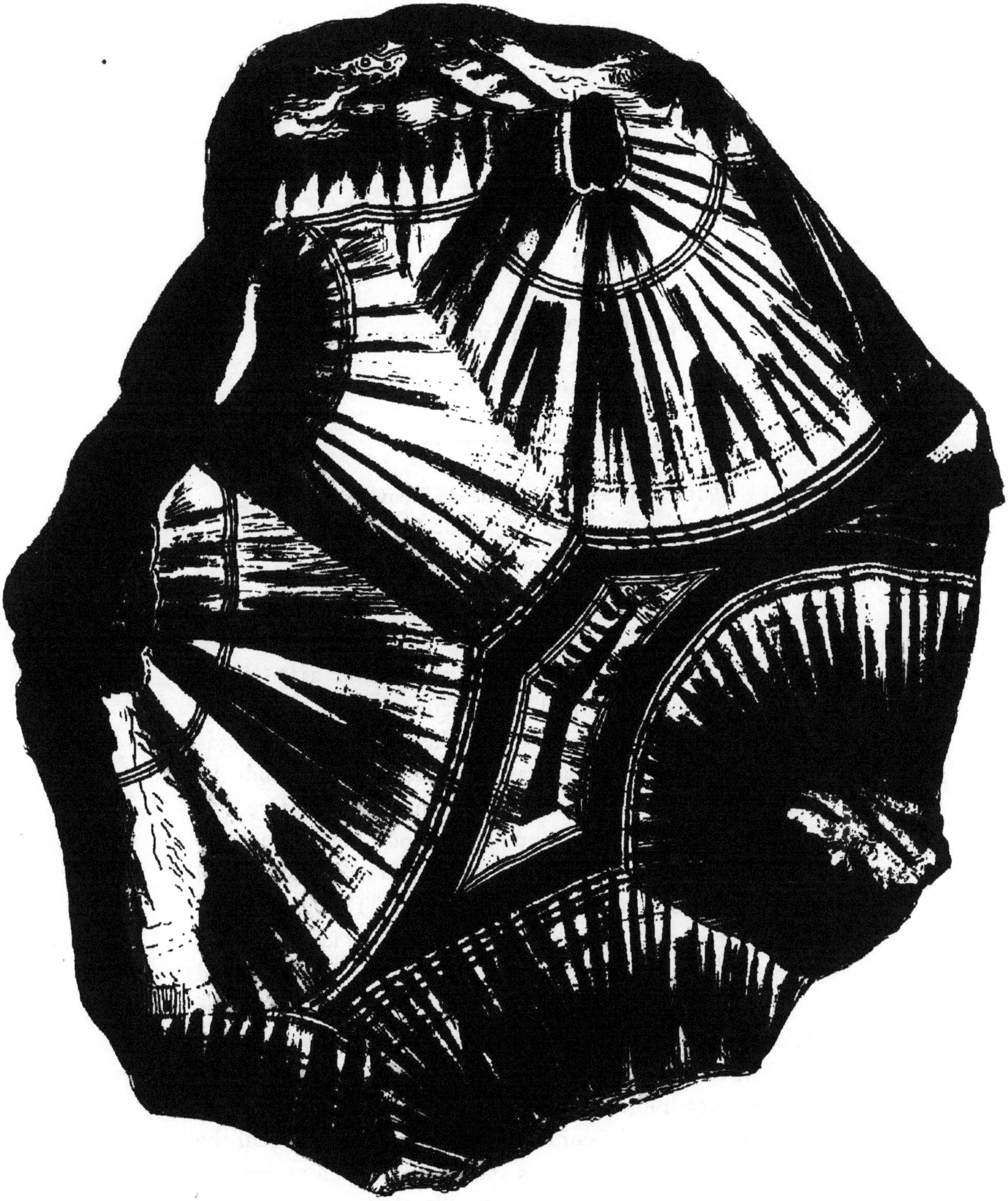

TAB. CXXXIII.

FERRUM oxygenizatum.

Wood-like Iron Ore.

Class 3. Metals. *Order* 1. Homogeneous.
Gen. 8. Iron. *Spec.* 3. Oxide of.
Div. 2. Imitative. *Var.* Wood-like.

CHARLES JOS. HARFORD, Esq., was so good as to bring the specimen from which this figure was taken, from Bristol. It was found in a ploughed field in that neighbourhood. It is of that variety often called Wood-like Iron Ore, and much resembles Wood-like Tin Ore. It however is much lighter and softer, and much larger than Wood-Tin has ever been found. Its weight is about 7lb., and its diameter about 6 inches by 5 inches. The figure being smaller, the many centres from which the radii form give it a pretty variegated appearance, and the radii are relieved by the darker tints of the striæ which are least ochry. The circles are so regular in some parts as to appear rather forced in the drawing: it may, however, be safely said that they are not more regular in the drawing than in the specimen. It may be curious to observe how reciprocal the force of formation is in this specimen, as each centre has terminated its radii with as little disturbance to its neighbour as possible. The primitive crystal of Oxide of Iron is not yet known; whatever it is, it must have some affinity to the Galena—*tab.* 131.

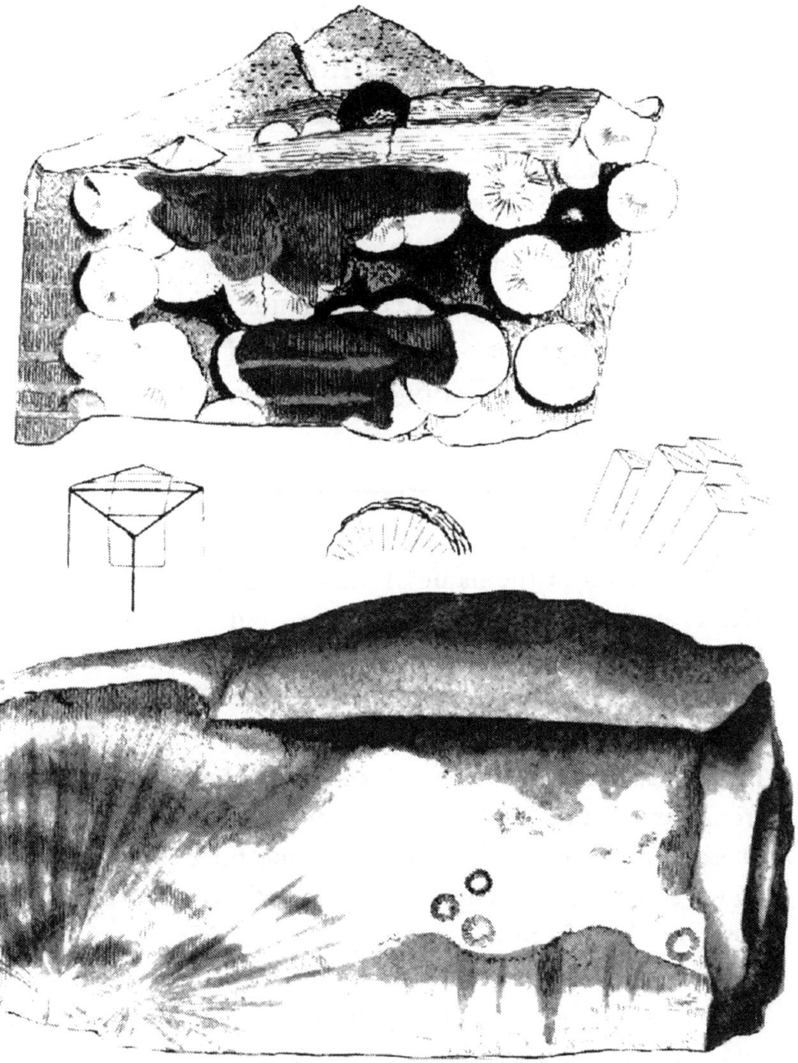

TAB. CXXXIV.

ARGILLA hydrata.
Hydrargillite.

Class 2. Earths. *Order* 1. Homogeneous.
Gen. 1. Argil. *Spec.* 2. Hydrargillite.
Div. 1. Crystallized.

SPEC. CHAR. Argil in combination with water.
SYN. Hydrargillite. *Davy in Phil. Trans.*
Wavellite. *Babington in Mr. Davy's Paper in Phil. Trans.*

ACCORDING to our system, we of necessity differ from Dr. Babington in the name of this mineral. We venerate Dr. Wavell as an old acquaintance, and have little doubt but his name will be handed down to posterity as the first discoverer of this substance, near Barnstaple. Through having in my possession the valuable collection of W. Day, Esq., I have had specimens of it several years under the title of a Zeolite from Devonshire, and was going to figure it as such; but on examining and finding it a very doubtful Zeolite, I put it by for some future analysis or investigation. It happened, however, luckily to fall into the hands of Mr. Davy, who fully investigated it, and from the nature of its composition he very properly names it Hydrargillite (from ὑδωρ water, and αργιλλος clay); and under so good an authority, we have little else to do than to quote a part of his paper from the Philosophical Transactions.

"The most common appearance is in hemisphærical groups of crystals" (on the surface of the gangue); "in some instances it exists as a collection of irregularly disposed prisms forming small veins in the stone: as yet I

(My specimens terminate the radii *outwardly*, something like Sulphate of Barytes, and seem to be parts of a depressed octaëdron, sometimes a little truncated: see *middle figures.*) "Its colour is white in a few cases with a tinge of gray or green, and in some pieces (apparently beginning to decompose) of yellow. Its lustre is silky, some nearly opaque. Its texture is loose." Our *upper figure* chiefly resembles this description. *The lower figure* has small dark circles, which seem to be the same substance with the margin darkest. Those larger spreading radii, which condense into white opaque circles with the help of smaller stellæ, terminate towards the ends.

The whiter parts are small opaque clusters, in which the three darker ones are imbedded.

A similar substance has been found near Truro in Cornwall, which has been examined by the Rev. W. Gregor, of which I hope hereafter to give a figure.

Mr. Davy's analysis of Hydrargillite:

Alumine . . .	70
Fluid	26·2
Lime	1·4
Loss	2·4
	100·0

Upon further examination we find it possible that the crystal may be derived from the cube, as we find the fracture perpendicular to the terminal faces of *the right hand figure* on a plane with the lengthened sides; and another parallel to the plane of the terminal face. I think I can with a magnifying lens discover a fracture on the sides. The column or prism is contrary to these faces, forming four sides, and there are two triangular faces set edgewise on the broader angle. We could not measure these faces, but hope that we may, some time or other, meet with larger

135

TAB. CXXXV.

PLUMBUM cupreo-antimoniatum sulphureum.

Sulphurated Cupreo-antimoniated Lead.

Class 3. Metals.	*Order* 1. Homogeneous.
Gen. 15. Lead.	*Spec.* 3. Sulphuret of Lead.

SYN. Triple Sulphuret of Lead, Antimony, and Copper. *Bournon and Hatchett in Phil. Trans.* 1804. *Part* I.

Ore of Antimony. *Rashleigh, v.* 1. *pl.* 19.

THIS comes from Huel Boys in the parish of Endellion, in Cornwall, and has been always rare, and was supposed to be a *Sulphuret of Antimony*, until Mr. Hatchett analysed a substance nearly related to it, but somewhat lighter-coloured, and found in the same mine—see *pl.* 136—which agrees with it in the nature of its crystallization, as is pointed out with much ingenuity by the Count de Bournon in the Philosophical Transactions.

They are found to be triple Sulphurets, and when most pure contain chiefly Lead, Antimony and Copper in the metallic state in union with Sulphur. Thus the whole becomes a cupreous antimoniated Galena, and, as most related to Lead Ore or Galena, we name it as above.

The present specimen is as it were passing into this triplet, as one of the common appearances of Sulphuret of

Antimony is with it in the form of hairs and spiculæ, and the crystals are composed of, as it were, bundles of fibres, making a more or less complete modification, and accumulating in whorls or in crosses: see *the right and left hand figures*. The signs of the primitive cube are distinctly seen upon the ends of the groups, and the reverse of the same figure; but it is difficult to see the primitive fracture, as, although they have a loose appearance, they are so incorporated that they fracture most like a compact glassy substance, conchoidally and irregularly. We however had the luck to find one fractured face which may agree with the Count's ideas. We have in this therefore the cubic sign of the primitive with the edges bevilled at an angle of 150° on the upper face, and of 120° on the prism, nearly as the Count de Bournon observes.

TAB. CXXXVI.

From the purity of these specimens, which were lent me by Mr. Richard Phillips, resulted a most perfect analysis by Mr. Hatchett, who has determined the proportion of the substances in it: and the largeness of the crystals has been the cause of Count de Bournon's determining the primitive crystal and modifications, which is so difficult, as he observes, from the irregularity of their increase. The primitive, as he remarks, is a rectangular tetraëdral prism, which has its terminal faces perpendicular to its axis. Thus it is a sort of cube, and from observation on my specimen (see *tab.* 135.) I find the integrant molecule may be a triëdral

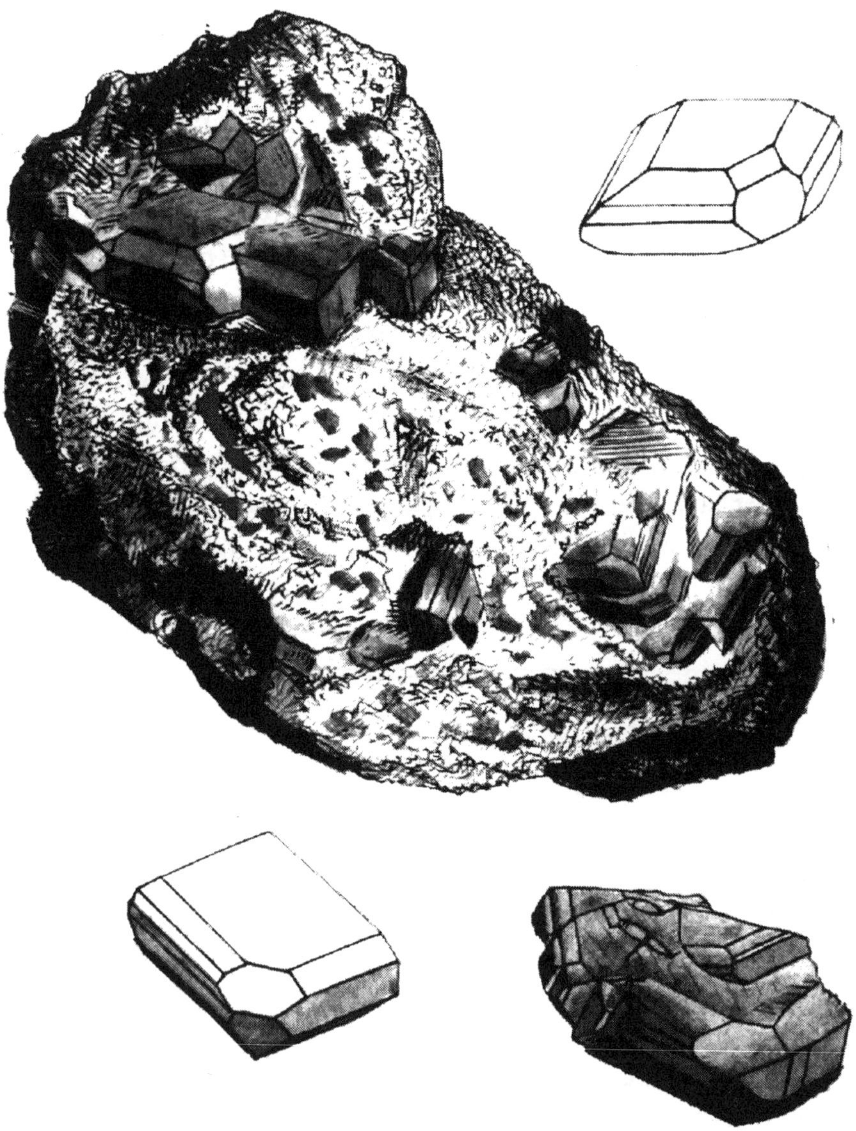

a

1

We have here figured what appears to be one of the largest crystals that have yet been seen: it is terminated at both ends with short columns in the form of plates, and stuck sideways on the gangue; the column forming hexaë-dral faces, chiefly on account of the deepness of the other faces, and the decrease towards the middle, mentioned at *tab.* 135. Thus the face decreasing on the column is reduced to a small triangle: see *right hand figure.* On this crystal we also observe another modification that has not yet been mentioned by Count de Bournon, viz. the upper face on the corner of what he calls the primitive prism, forming from the terminal face probably at the same angle with those on a line with the prism: see figure *a.*

Mr. Hatchett, after a careful analysis, found it to contain:

Lead . . .	42·62
Antimony .	24·23
Sulphur . .	17
Copper . .	12·80
Iron . . .	1·20
Loss . . .	2·15
	100·00

It is of a grayer colour, and much more fusible than Galena, as it melts before it is red hot; it leaves a cupreous residuum, whereas the fibrous part—*tab.* 135—leaves scarce any. Its lustre is very great. *Fig.* 1 points out the faces of the column, which are very small. *The right hand figure* is the natural crystal; and *the left hand figure* is a geometrical outline, to explain the faces above and below the column.

TAB. CXXXVII.

HYDROGEN Bitumen.

Elastic Bitumen, or Fossil Caout-chou.

Class 1. Combustibles. *Order* 1. Homogeneous.
Gen. 1. Hydrogen*. *Spec.* 1. Bitumen.

GEN. CHAR. Inflammable, easily converted into gas by calor. Forms water by combustion with ogygen gas.

SPEC. CHAR. Nearly pure, fœtid, not easily volatile.

SYN. Elastic Bitumen. *Hatchett in Linn. Trans.* *v.* 4. 146.
Bitume élastique. *Haüy*, *v.* 3. 313.
Elastic Bitumen. *Schmeisser*, *v.* 1. 290.
Mineral Cahoutchou. *Kirwan*, *v.* 2. 48.
Elastisches erdpech. *Karsten*, 42.
Cahoutchou fossile. *Lamétherie*, *v.* 2. 540.

THE Elastic Bitumens were first noticed at Castleton, in Derbyshire, about the year 1786. Perhaps their general resemblance to the Caout-chou, or Indian rubber, discovered about half a century ago, might in a great measure be the cause of their being noticed; for we are often struck with the comparative resemblance of a thing which other-

* Unknown in its pure state, unless as the softest and purest Bitumen.

wise might have passed unnoticed. It is however a curious circumstance that they have not been discovered elsewhere, although Petroleum, Naphta, and analogous substances, as Maltha, Mineral Tar, Pitch, and Asphaltum, which are nearly related to the above, are found in many parts of the world, as well as at Castleton, where this substance is found. These always differ from the vegetable substances of the same nature (viz.) common Tar and Pitch, by their peculiar odour, which somewhat resembles oil of brick, used by Lapidaries, and which I understand is a kind of burnt oil. It is very different from the well known scent of Pitch and Tar. Vegetable Caout-chou differs from both. So, we may say that the under-ground change seems to be the cause of the odour. We cannot at present account for the elasticity, otherwise than in the words of the ingenious Mr. Hatchett in the Linnean Transactions: "From what I have already related, I suspect that the elastic property is occasioned by the interposition of very minute portions of air, or some other elastic fluid between the parts of the Bitumen, and that this takes place by means of some unknown cause at the time of formation; for, when these Bitumens are melted, the elastic fluid is liberated, and the mass loses that fine spongy texture which I suspect to have been the cause of the elastic property."

It is somewhat curious that it is found as it were oozing out of, or attached to, the rocks. The present specimen is attached to common Limestone rock, mostly Stinkstein: see *p.* 81 *Brit. Min.* It is nearly the softest of the elastic sort: some parts of it are almost in an oleaginous state, and stick to the fingers. It is nearly the colour of common Caout-chou, but will not stretch out like it, although it springs to its form after compression; and hence it differs a little from Caout-chou, as the indenture of the impression is not so easily made permanent. It is by no means so tough as common Indian rubber.

Sept. 1 1806 Publish'd by Jas. Sowerby London

TAB. CXXXVIII.

This is a harder specimen than the former, and much about the density of common Caout-chou. It is somewhat remarkable for its peculiar conchoidal fracture, on the faces of which, when minutely examined, peculiar risings occur, bounded by curved lines crossing each other, and terminating in a kind of centre not unaptly resembling the fracture of Carbonate of Lime with a kind of curvilinear crystallization*; the cracks seem not at all guided by this. The elasticity is something like the former, as it soon separates if we endeavour to stretch it out. It is externally of a blooming gray, and internally of an olive green colour. This is oozing more or less from different parts of the gangue, which is a mixture of Carbonate of Lime, Galena, &c. Some of the smaller particles have a reddish illinition in their flaws.

* I have lately received a fine specimen of Carbonate of Lime from Mr. Hall of Arkendale, very expressive of this.

1

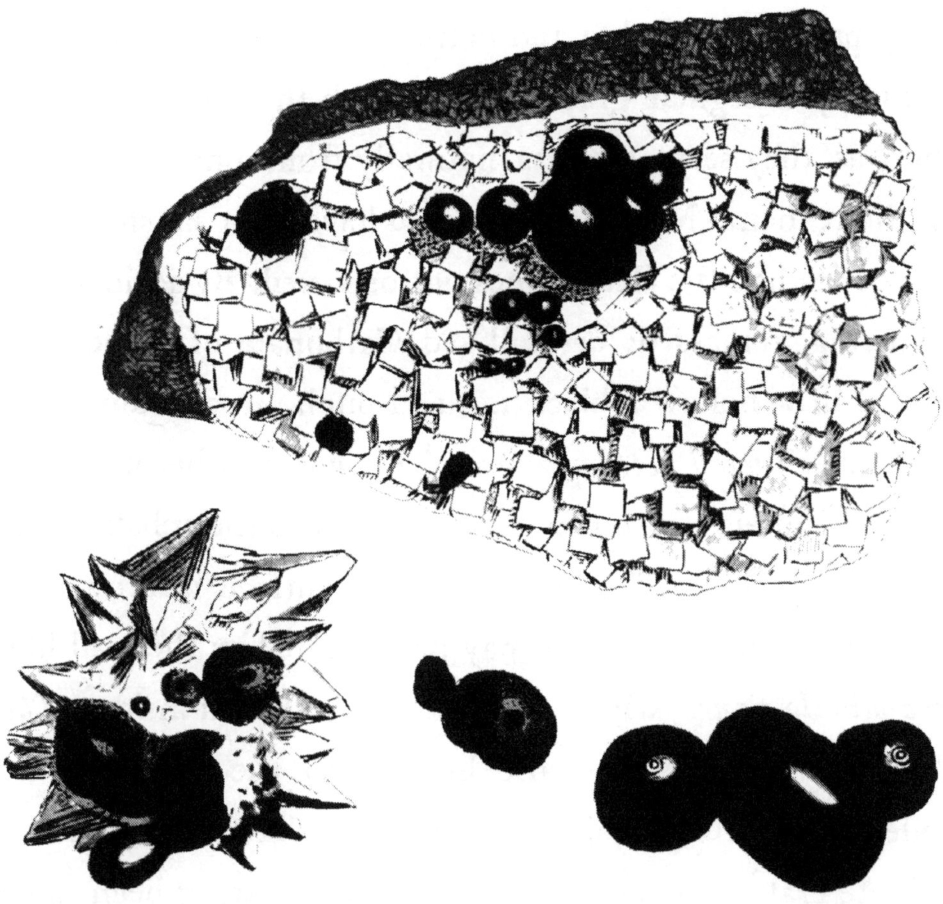

In examining the Bitumens, it is difficult to say whether they pass from Naphta and Petroleum to Pitch and Asphaltum, by a regular gradation, through the elastic kind, or not, as they seem to pass naturally from one to another without them. Most people would have been satisfied with the series had the elastic ones not occurred. By accidentally breaking a mass of crystallized Carbonate of Lime, was found in a hollow a black mineral pitch, in a liquid state: see *the left hand bottom figure*. This has now become condensed and elastic, but not so much so as the substance in the last figure. The crust or outer surface is brownish, with more elasticity, and may be separated by the nail like *the middle figure*, which shows the outside and inside. *The left hand figure* shows also hollows in the centre of the outer crust, or external parts, something like the mouth of a minute crater; giving a strong idea of its having been once in a state of powerful ebullition from that hole. This is a darker-coloured Bitumen, possessed of greater elasticity than any that has been before noticed. Among specimens this should be placed as the darkest, and perhaps nearly the last of the elastic sorts. The next in this plate is a more indurated the circular indentures remaining on the bubbles: see *the right hand figures at the bottom.* They are very neatly formed upon whitish cubic Fluor, and seem as if they had splashed about in falling; for they give the idea of having dropped on the surface of the crystals of the Fluor in a state of strong ebullition, perhaps more so than the former. Whatever is the cause, the effect may be gained by a stronger heat; as, the nearer these substances approach combustion, the more they harden, and form the appearance of Asphaltum, which we suspect this substance to be. This is from the samé neighbourhood as the others, and is totally black all through. Fracture conchoidal and shining, destitute of any lighter illinitions, being perfectly opaque.

TAB. CXL.

Of the indurated Bitumens, one of the first we met with was included in coarse, somewhat earthy, elastic Bitumen, and much represented in form a gun flint, with fractures not unlike those in common flints. It proved extremely pure, of a dark olive green, and was closely surrounded by the other bitumen, as if it had been the fragment of a larger piece preceding in formation the earthy sort. We have since found a larger specimen of a similar nature, with a black outside—see *the bottom figure*, and also a small stalactite or lengthened drop. The smaller fractures in these show transmitted lights or illinitions of a warm yellow colour.

Thus it should appear that an accumulation of this yellow light, mixed with the black, gives the greenish hue to the substance. We suspect that these colours are more or less caused by the different degrees of oxygenizement of Iron.

Dull greens are seldom happily imitated by illuminators; thus, our figures may in this instance be rather too gay. According to the best chemists, Bitumen is formed principally of Hydrogen* with more or less Oxide of Carbon, Oxide of Iron, and other incidental substances.

* Combined with but a small portion of Caler

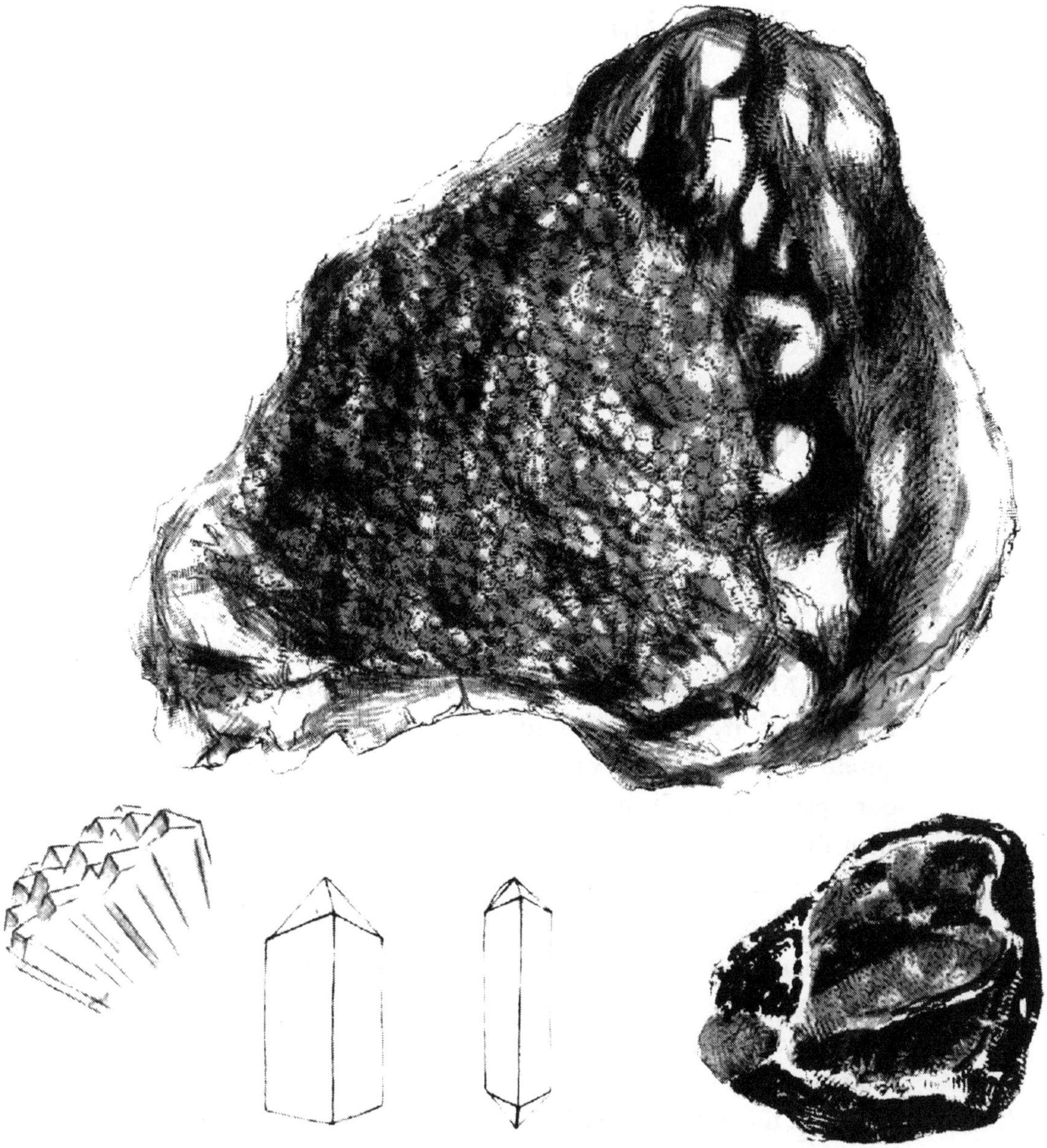

TAB. CXLI.

FERRUM Cupreo-arseniatum.

Cupreous Arseniate of Iron.

Class 3. Metals. *Order* 1. Homogeneous.
Gen. 7. Iron. *Spec.* Arseniate.
Var. Cupreous.

SPEC. CHAR. Iron, Copper, and Arsenic Acid in combination.

SYN. Cupreous Arseniate of Iron. *Bournon and Chenevix, Phil. Trans. for* 1801.

THIS substance, which is rare in Cornwall, we understand has been brought from Siberia by Professor Pallas, according to Mr. Chenevix, to whom we are obliged for the analysis of it, and to Count Bournon for the most accurate account of its crystallization, which agrees with our specimens. It is found in the Muttrel mine along with the cubic Arseniate of Iron. The crystals are always small, generally clustered, more or less in bundles or confused. The individuals form rhomboidal prisms having the two opposite angles very obtuse; and consequently the two others are very acute, terminating with four scalene triangular faces, fixed lower on the acute angles than on the obtuse ones. We cannot be certain that the angle bearing upon the obtuse side of the pyramid is not a right angle: see *the geometrical figure. The upper specimen* is a largish collection of clusters variously grouped, the clusters formed of crystals diverging from a centre, with the faces of the pyramid only exposed. It is on an irregular gangue of white Quartz with some blackish lumps of gray Sulphuret of Copper, and a few rectangular plates, perhaps Uranite. *The left hand bottom figure* represents a specimen with more distinct crystals looking like spiculæ, scattered in the hollows of an ochraceous gangue. These sometimes expose one, and at others both ends: see *the lower figure*. These crystals are mostly of a light shining sage green.

Analysis by Mr. Chenevix:

Silica	03
Arsenic Acid . .	33·5
Oxide of Iron . .	27·5
Oxide of Copper .	22·5
Water	12

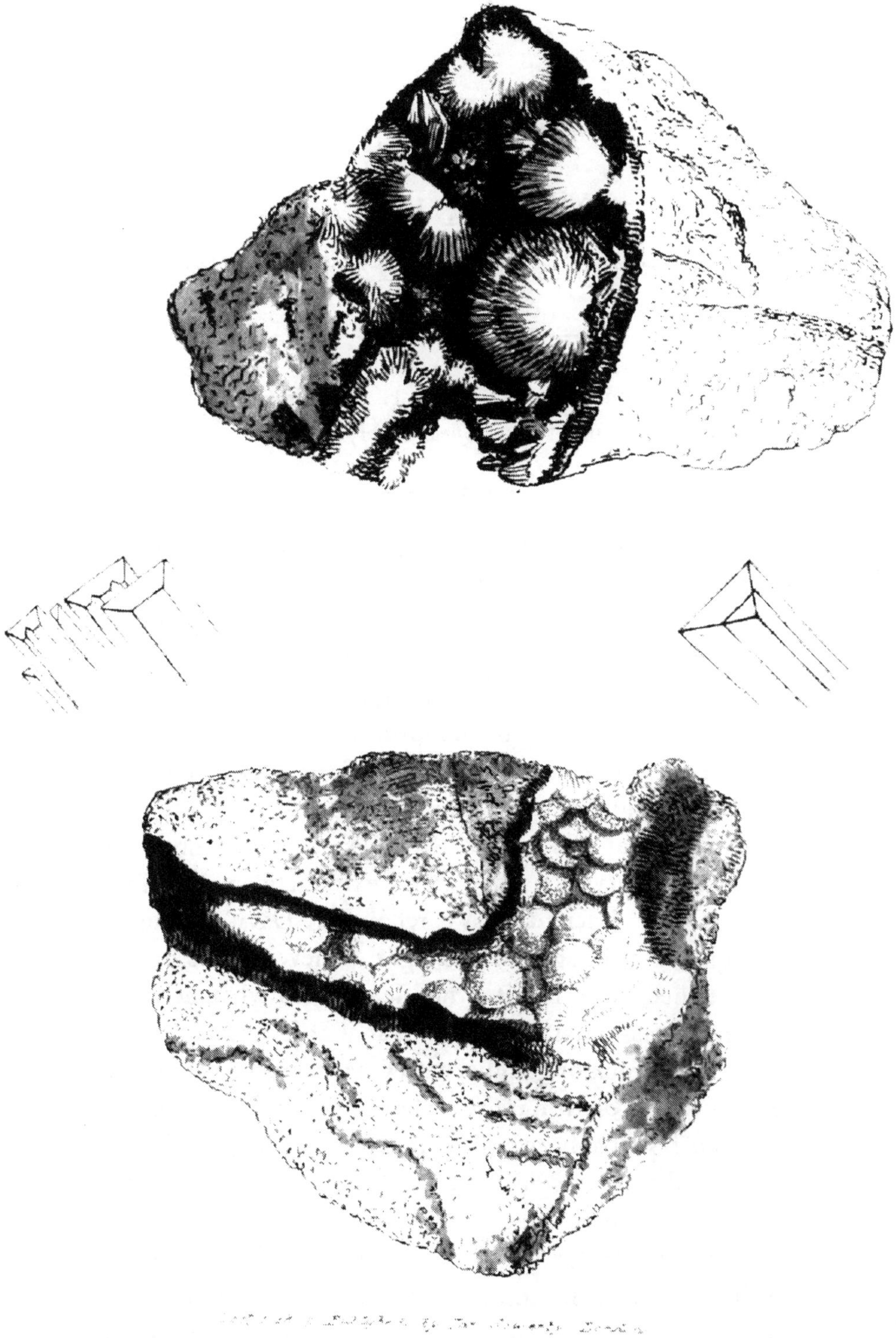

TAB. CXLII.

ARGILLA hydrata.

Hydrate of Argill, or Hydrargillite.

Class 2. Earths. *Order* 1. Homogeneous.
Gen. 1. Argill. *Spec.* 2. Hydrate.

According to our expectation before expressed, we are enabled to give figures from Cornish specimens of this substance by favour of the Rev. Mr. Gregor, who lent us his best specimen, which is represented in *the upper figure*. This is in looser radii than those from Barnstaple, but appears to agree precisely in the form of the crystal, which, according to our idea, shows more of the primitive faces, (viz.) the flat sides of the column, as that shows only one primitive face at the apex. We could not attempt to measure these crystals, they are so extremely small. The gangue is also different. We retain Mr. Davy's name of Hydrargillite, although Mr. Gregor believes there is an acid contained in it, but it is not determined what acid, or whether it is essential to the mineral. It is formed in the hollows of the rock or gangue, with Quartz crystals coloured with red Oxide of Iron more or less sticking in mammillæ about them, probably mixed with Oxide of Copper; but it has not undergone regular examination. On some parts of the gangue there appears to be Oxide of Uranite: see *tab.* 124; and Mr. Gregor, by some trials made with small quantities, thinks it contains, beside Uranium, a little

Oxide of Lead, Lime, Silica, and a metallic substance differing in properties from Uranium.

We hope this will be further examined. Mr. G.'s paper on this subject to the Royal Society has been read, but is not yet printed. We have another specimen from our friend Mr. Rashleigh, where the spiculæ are much smaller, and the bundles composed in such a manner as to give it a cottony appearance, whilst a few of the crystals are nearly like the above.

We have annexed to the plate a specimen from our liberal friend last mentioned, which accords extremely well with the common appearance of that from Barnstaple, showing them to be the same, which has been doubted, and the crystallization sufficiently corresponds with them. The part of the gangue in which this is imbedded is of a grayish colour, and somewhat though not distinctly approaching a schistose appearance, like those from Barnstaple. The rest of the matrix resembles that of the others from Cornwall, which are whitish Quartz mingled with soft Mica, such as the Cornish Apatite is generally found in. Those from Cornwall were supposed to be different species from those from Barnstaple. We are more confident with regard to the crystals, from having been favoured by Dr. Wavell with many interesting specimens, some of which have a pearly appearance not before mentioned. All the specimens figured in the present plate are from Stenna Gwinn in Cornwall.

143

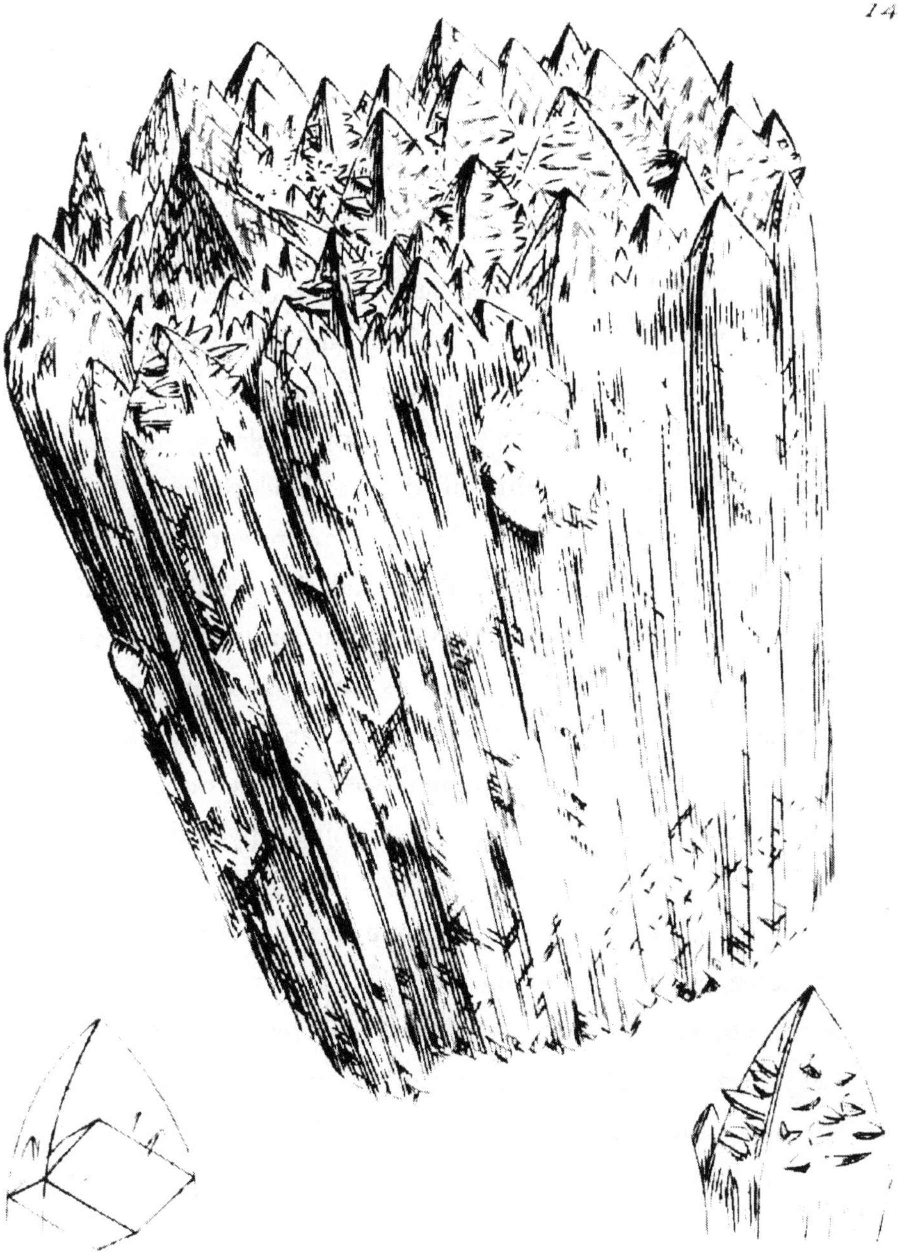

TAB. CXLIII.

CALX carbonata; *var.* inversa.

Inverse crystallized Carbonate of Lime.

Carbonate of Lime has been remarked at Portland Island for its fine topazine colour, and has thence been denominated Sugar-candied Spar, or Candied Sugar Spar. These crystals are not uncommon in other places, but are often finest in the fissures of the Portland Stone. The Stalagmites from Bath, Yorkshire, &c. present the same substance, generally less massy, and the crystallization more or less in spiculæ, which often accord with this in form: therefore this will serve as an illustration of all, whether small or large, and most of the fanciful formations are of this nature. The present specimen terminates in acute rhombs somewhat rounded, forming a pyramid upon the obtuse angle of the nucleus: see *the left hand geometrical figure.* These often have some crystals formed under the same circumstances on the sides; and it is a curiosity worth observation here, that these smaller ones are probably formed at a time when the others are nearly perfected but not quite so, as the crystallization is somewhat independent; yet they were enabled to stick so into the larger ones, as to be of equal solidity with the rest of the mass.

TAB. CXLIV.

This is nearly of the same formation as the last, the crystallization radiating from various centres with the terminations meeting*: they however are in more determined columns of three sides, and the pyramids at first sight appear more simple: however, there is a tendency to double each face of the pyramid, making, like the metastatique, a six-sided pyramid. This is one of the darkest-coloured ones of this nature.

* The Radii in Mr. Hall's specimen, mentioned *tab.* 138, are a foot long, the sides of which are divided by a double direction of curved lines crossing to a central bar, consequently circumscribing rhombic forms with two convex and two concave sides.

TAB. CXLV.

CUPRUM oxygenizatum.

Crystallized Red Oxide of Copper.

Div. 1. Crystallized.

I HAVE already figured some Cubical and Octaëdral Ruby Coppers. The present one from Cornwall is extremely rare and curious, as it has a remarkable modification—the cubo-octaëdron of Haüy, with the addition of an obtuse four-sided pyramid on each cubic face, formed as it were of somewhat distinct plates. In one crystal this pyramid is formed into a cross, the plates of which it is composed being notched or incomplete in the centre of their edges, and complete at the corners: see *the right hand upper figure*. With these are crystals of similar forms, of very pure Copper, composed of oblong particles with a reticulated appearance, as if they were some of the others, of an anterior formation, having been deprived of their oxygen. These crystals are very extraordinary, as they comprehend the Octaëdron with truncated edges: they have also truncated and bevelled solid angles, making a very compound crystal: see *the left hand figure*. The intermediate varieties are the dodecaëdron and octaëdron, truncated at the solid angles. The particles show some signs of being thread-like confused octaëdrons, somewhat resembling those on the following plate, *tab.* 147. They are externally between a copper colour and red, scarcely metallic in the lustre, until cut, and then perfectly so. It is, as the Count de Bournon observes, useful in many instances to use a magnifier to examine crystals; and we should lose some very interesting beauties without it. These indeed may be pretty well seen by the eye alone, but it is admirable to observe how neatly these are formed by such help as the magnifying glass. We do not know of any other specimen of this kind.

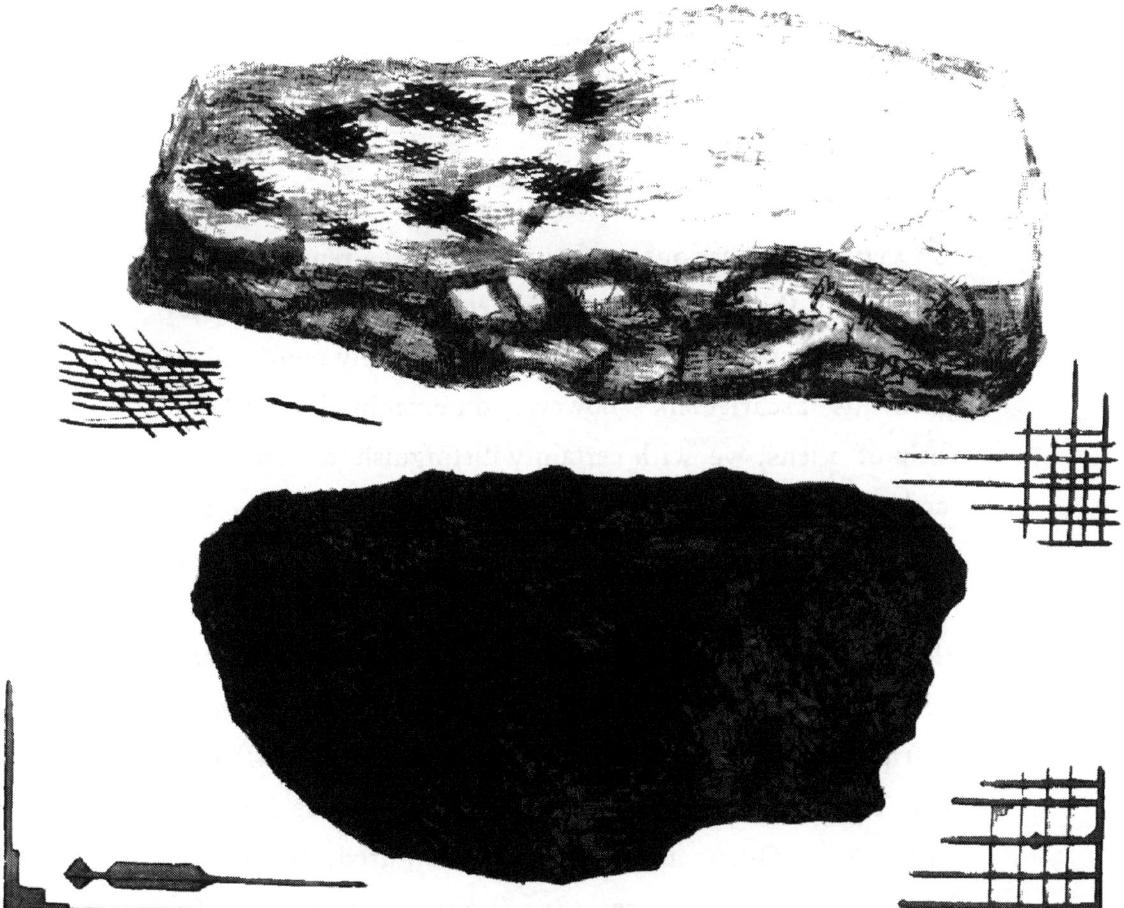

TAB. CXLVII.

CUPRUM oxygenizatum.

Filamentose Oxide of Copper.

Div. 2. Imitative.

NATURE, ever various and instructive, often represents one thing with the appearance of another. Thus, a casual observer would expect that this Copper Ore merely consisted of filaments of scarlet silk: however, on examination with the help of a lens, we with certainty distinguish the contrary, and see how beautifully Nature varies her operations, and under certain circumstances accomplishes the same end with different appearances: thus, the moleculæ are forming threads more or less discoverable in the shape of elongated octaëdrons or cubes. *The upper figure* is decomposing Feltspar and Quartz with these filaments of Ruby Copper Ore, some of which are in bent 4-sided threads crossing each other. Others are irregular and confused, apparently having been disturbed: see *the left hand upper figure.* In some parts they are disposed in straight filaments, crossing each other at right angles, as if disposed to form a cube: see *the right hand upper figure. The lower figure* is chiefly Ruby

Copper and Native Copper, with a little Quartz and Copper Pyrites. The fibres on this are larger, and show signs of elongated octaëdrons and rectangular prisms. They are beautifully pellucid and transparent, with the full lustre of a ruby. We may add that there is a regular gradation from the powdery oxide, naturally opaque, through masses of confused filaments so extremely fine as to partake of the same opacity, to such as resemble fine wool.

We were favoured with this last specimen by Messrs. W. and R. Phillips, who had it from Redruth in Cornwall.

1
2
3
4
4
3
2
1

TAB. CXLVII.

CALX carbonata, dura.

Hard Carbonate of Lime

Class 2. Earths. *Order* 1. Homogeneous.
Gen. 3. Calx. *Spec.* 5. Carbonate of Lime.
Div. 1. Crystallized*.

SYN. A new species of hard Carbonate of Lime. *Bournon, Phil. Trans.* 1803. 325.

GREAT BRITAIN is highly indebted to the Right Hon. Charles Greville's mineralogical ardour, for one of the most complete and scientific collections in the world; and I am happy to congratulate my friends on his generosity in readily allowing me the free use of it, as nothing can now be wanted, as far as is at present known, to complete the British history of this kingdom of Nature, with my own collection and those of other friends already mentioned.

The present specimen from Scotland is a great curiosity to the mineralogical world, and was first noticed in Mr. Greville's collection by the Count de Bournon, who has given an excellent account of it in *Phil. Trans.*

We understand that only a few specimens have been preserved, which were collected in the neighbourhood of Glasgow.

The structure at first sight has something new in it; but it might be taken for Carbonate of Lime, which it really is, although the fracture much resembles that of Quartz, but is somewhat rougher. We admire the Count de Bournon's indefatigable patience in measuring and pointing out the modification of these crystals, which we did not attempt; we have only ventured to point out a few faces that were passed over, as they may possibly be somewhat interesting; for which reason we have made sketches of them. The bases seem exactly as the Count has determined them. The principal are 3 small ones on the summit, as represented *fig.* 2.—*Fig.* 4. shows a minute triangular one, and one of those above mentioned. The matrix is ferriferous Carbonate of Lime with Pyrites and small double-pointed crystals of Quartz.

Flos-ferri and Arragon Spar of old authors belong to this variety.

TAB. CXLVIII.

CALX carbonata fœtida.

Ochraceous Stinkstein.

Div. 2. Imitative.

The formation of this substance, however singular, seems hitherto to have escaped notice. It may nevertheless lead us to a determination in many cases. It might at first be taken for a Coralline; but we have by comparison of specimens convinced ourselves that it is rather an assemblage of funnel-shaped Stalactites formed in a fluid medium, the surface of which has become encrusted at regular intervals, especially around the Stalactite. Although there is a variety of specimens, yet the structure coincides very accurately in many of them. Some indeed are more puzzling to account for than the present. It not uncommonly happens that Stalactites are hollow, (see *tab.* 6.) and others undulated. They also evidently form a deposit, or case after case, on the outside in a concentric manner. This does not seem to have been formed so; the peculiar state of the substance of which it appears to have been composed, having only a certain quantity of moisture, enough to form a kind of paste, which may have allowed it to have dropped into one mass at more or less regular periods, producing this remarkable appearance. Now, it happens that the spot which produces a variety of these produces also the Botryoidal Stinkstein in great abundance and variety: see *tab.* 38. They are generally found filled with a dusty ochraceous marle, such as would readily allow scattered drops of water to collect it on their surfaces.

I received most of my specimens from my friend N. J. Winch, Esq., who collected them near Sunderland.

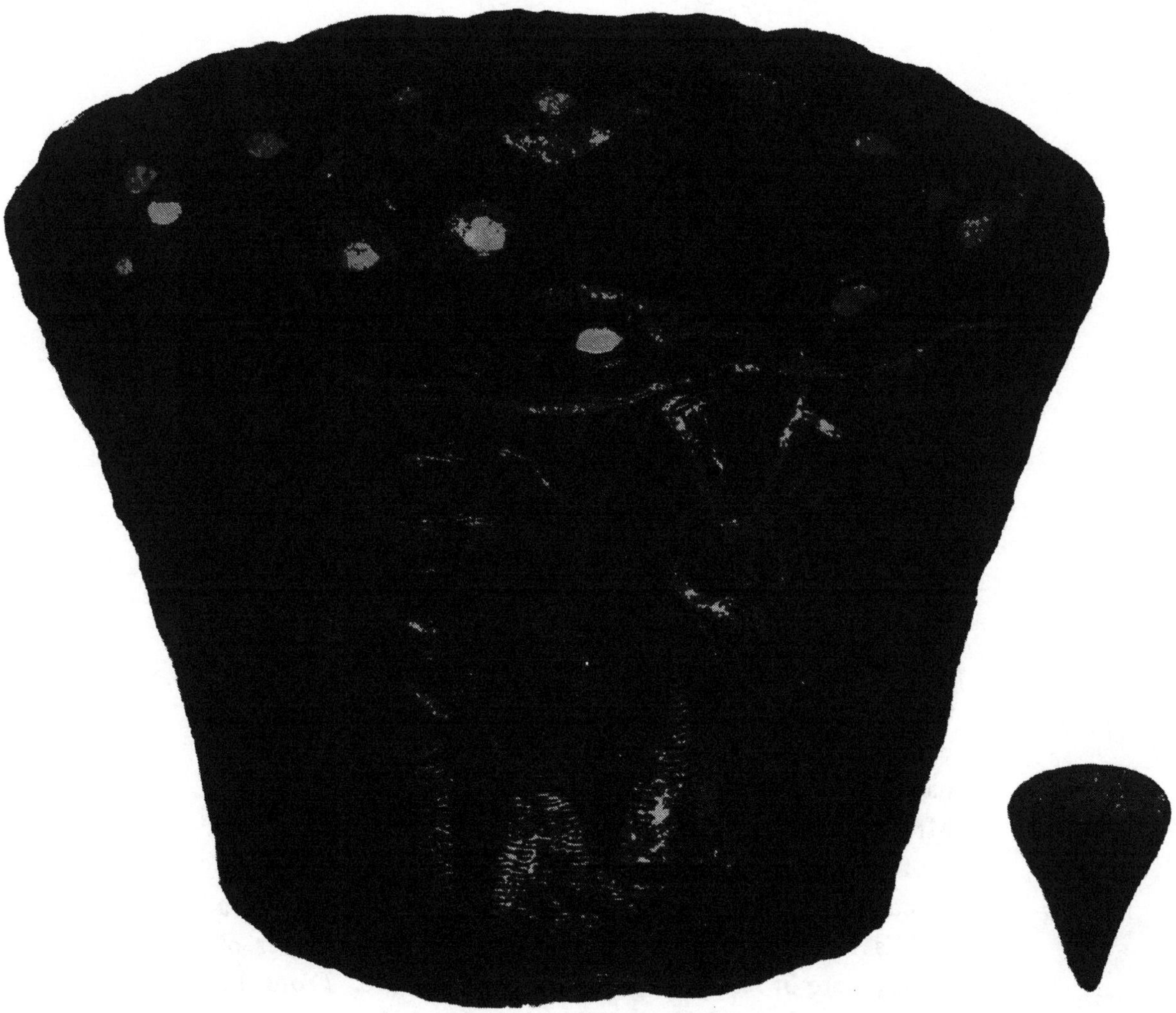

TAB. CXLIX.

ARGILLA marga.
Argillaceous Marle.

Class 2. Earths. *Order* 2. Compound.
Gen. 1. Argilla. *Spec.* 1. Marle.
Div. 1. Imitative.

SYN. Tophus turbinatus. *Linn. Syst. Nat. ed.* 13. *v.* 3. 190.

WE so frequently find the figure of Corallines, or other organized substances, that we are often puzzled to account for them; some however are readily understood to be infiltrations taking place of them, and that before us has hitherto been considered by many in this light. The singular regularity of the specimen here figured has given rise to many conjectures. In giving my opinion I desire to discover the truth; and having made many comparisons, I agree with Linnæus, that it is rather a Stalactite formed under certain circumstances among other operations of nature, which may be continued to a great extent. Thus, Mr. Martyn observes that a very large space in Derbyshire is of this formation. It is finely undulated: in other respects the cones run into each other, something like a child's horn gig—which will separate into a number of cones, more or less perfect, if placed by the fire. From the present specimen I separated some cones, by alternately wetting and drying it. We do not know how minutely they may be divided. The fracture is like other compact calcareous marles, that are not governed by the conical formation. We have had them from two or three friends in Derbyshire. The present specimen was given me by Mr. Baker, by favour of my friend the Rev. James Dalton, and comes from the alum-works at Boulby in Yorkshire belonging to the former gentleman, and was remarked for being in the form of a horse's hoof, having settled upon a Cornu-Ammonis.

They are mostly of an argillaceous marle: but I have one from Cumberland, given me by Mr. Buchanan from Barton fell, which seems to be more of an Iron ore than any of my others. In most of the English specimens the cones are rather confusedly coalesced. In foreign ones, groups of cones joined together by their edges are apt to separate; whence Linnæus's description. The present specimen figured contains so much pyrites dispersed through it, that it might be used as an alum ore.

TAB. CL.

ARSENICUM ferreum.

Ironey Arsenic.

Class 3. Metals. *Order* 1. Homogeneous.
Gen. Arsenic. *Spec.* Combined with Iron.
Div. 1. Crystallized.

SPEC. CHAR. Arsenic in combination with Iron.
SYN. Fer arsenical. *Haüy*, 4. 56.
Arsenic pyriteux. Mispickel. *Born*, 2. 197.
Native Arsenic alloyed with Iron. *Kirw.* 2. 256.

THIS substance much resembles the whitish Iron Pyrites or arsenical Sulphuret of Iron, and seems to have been partly confounded with it by Kirwan; who observes, that "the Marcasite found near Dublin, called Irish Diamonds, is of this species." This latter is however to be most readily distinguished by the crystals being right-angled*, whereas the present is an acutely rhomboidal prism. It is perhaps not unnatural for Irish Pyrites to have a little arsenic, but not in sufficient quantity to characterize a species. The colour is different in the interior; one being a grayish white, the other yellowish.

* A small portion of arsenic may be found in them, but not sufficient to alter the crystal.

The specimen figured came from Cornwall, and is remarkable for having somewhat acicular crystals collected in divaricating bunches, which pervade Blende, Copper Pyrites, and Quartz. Before the blowpipe it melts easily, gives out copious white fumes strongly scented with Garlic, and highly noxious to the lungs: what remains is a small gray globule of magnetic Iron.

The right hand figure shows the primitive rhomboidal termination of an elongated crystal, with a decrement upon the obtuse solid angle. The face produced by this decrement gives for its incidence upon the rhomboidal face 149° or thereabouts: this modification seems to be new, as Haüy has not mentioned it.

The left hand figure represents another view of the primitive, with a decrement on the acute angle, which agrees with letter *s* of Haüy's *fig.* 137.

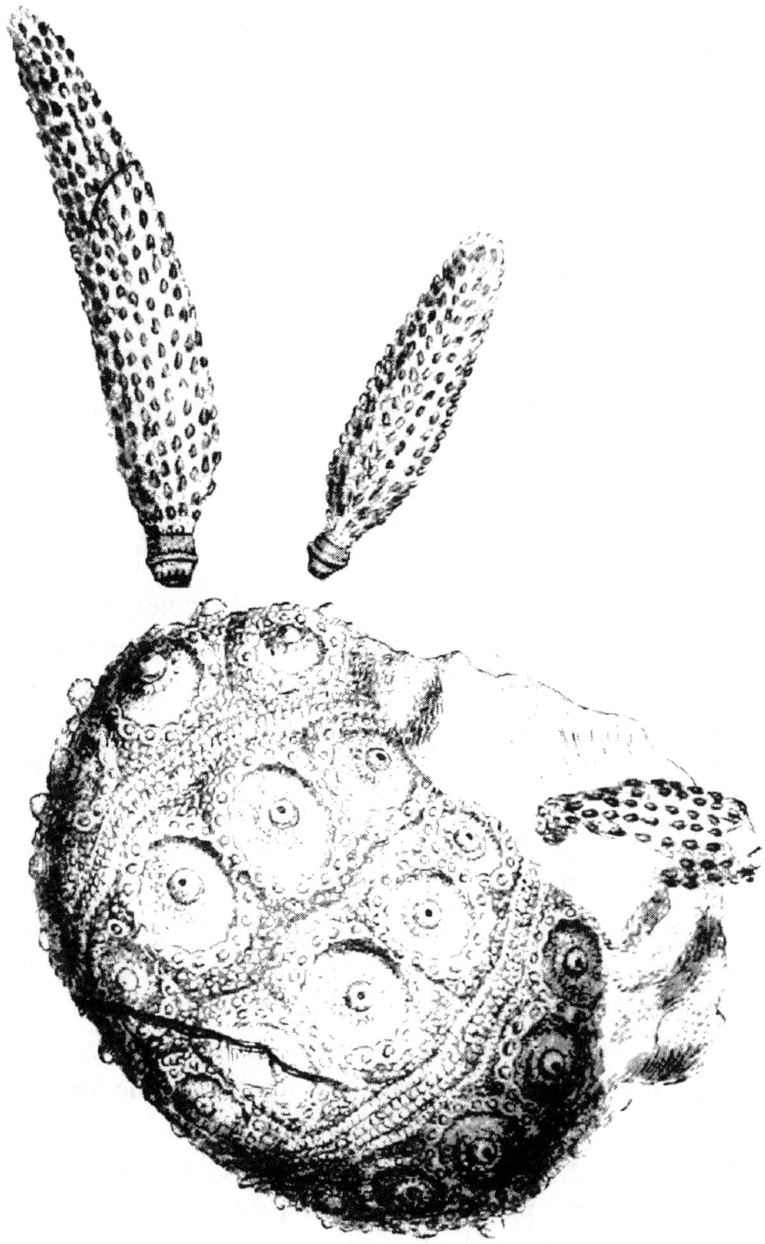

TAB. CLI.

CALX carbonata, echiniformis.
Echinus-formed Carbonate of Lime.

Class 2. Earths. *Order* 1. Homogeneous.
Gen. 3. Lime. *Spec.* 5. Carbonate of Lime.
Div. 2. Imitative.

THAT animals are indebted to minerals is very evident, although vegetables may be the instruments by which their nutriment is prepared. We may also say that great part of the mineral world is much indebted to the animal for its present appearance; for, under certain circumstances, nature allows the animal construction to remain long after the animal itself. In this instance, it might have been a doubt whether a construction so complete as this *Echinus* and its spines could be an infiltration of Carbonate of Lime in place of the case, or rather bone, or the remains of the Carbonate of Lime which existed while the animal possessed it. Its fracture is sufficient to determine it to be Carbonate of Lime, without any other trial; and upon examination it is found to be nearly pure.

Mr. Hatchett, in his valuable account of the shells and bones of animals, proves the case of a recent *Echinus* to be bone, as it contains its due proportion of Phosphate of Lime. This is therefore the more curious, as there is no Phosphorus remaining to destroy the crystallized character; the carbonic acid predominating effectually in that particular. To know the animals that are thus found as it were petrified may not seem at all the province of the mineralogist, nor is it perhaps strictly necessary; yet it is convenient to have such information, as it may be sometimes of great importance; and it must appear remarkable to all, that although the petrified remains of other animals are very universal, yet those of our own species are never found. The effects of the change in the *Echinus* are remarkable, as it appears to be extremely gentle; for the parts of the shells which are of a delicate structure, and are easily separated in common decay, are often very little disturbed by the changes of the animal substance. The animal parts, exclusive of the Carbonate of Lime, must pass away, while Carbonate of Lime, or Flint, is filling up the vacancy.

This species of *Echinus* is either a new one, or a variety of *Echinus Cidaris* of *British Miscellany*, and was sent to me from Queensford by my kind friend T. Mead, Esq. The soil in which it was found is a chalky marle.

TAB. CLII.

THE apparent loss of the animals in these *Echini* is as yet unaccounted for, and an immense quantity must have been lost to our view; although, on the examination or analysis of Earths, the animal substances have never been discovered. Mineralogists have indeed lately discovered *alkali* in some Basalts, &c. which was always supposed to belong to the vegetable kingdom. This specimen was found in a chalky

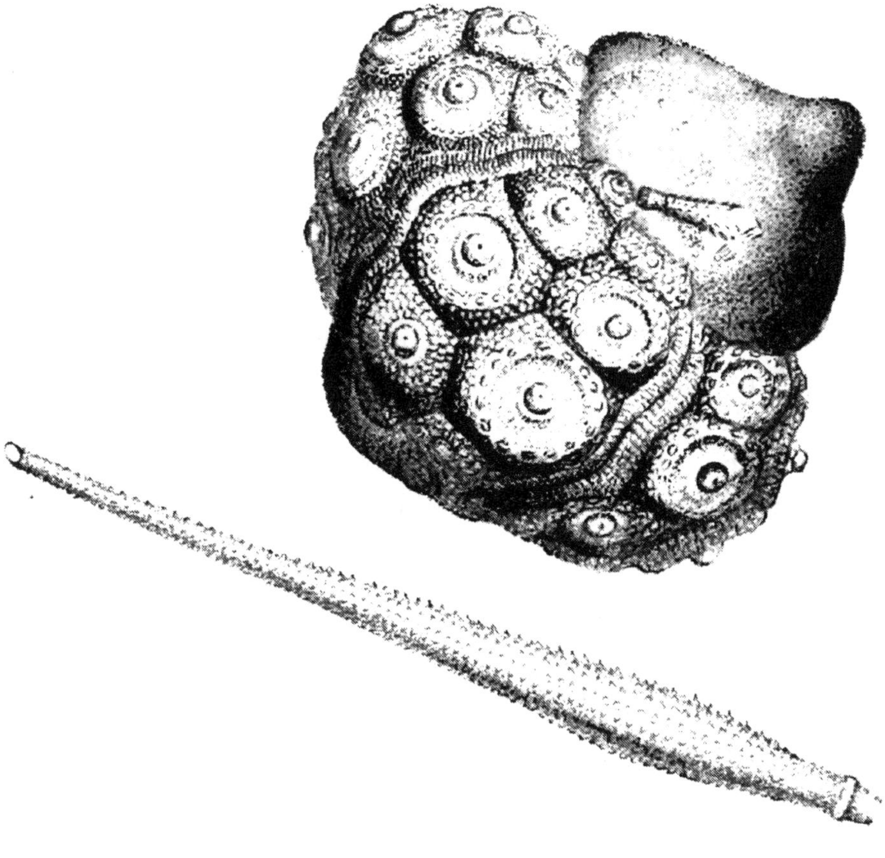

Dec.r 1 1805. Published by Jas. Sowerby London

rock at Saffron Walden by my friend Thomas Walford, Esq. We have the spines from various places. Its great curiosity is, that the shell part is extremely perfect as a Carbonate of Lime, although filled up with Flint; which has scarcely disturbed it, notwithstanding its having overflowed as it were at both ends. This is also said to be a variety of the *Echinus Cidaris;* to which we do not readily assent. P. Rashleigh, Esq. has figured another species, *Echinus circinatus,* under similar circumstances; which not being uncommon, serves well to show that nature performs the same offices by similar means in various places; and, as Mr. Rashleigh observes, its particular structure will point out to the theorist that it is not the immediate effect of *fire,* in the common acceptation of that term, but, with the aid of other principles, modifying the operation so as to give another idea.

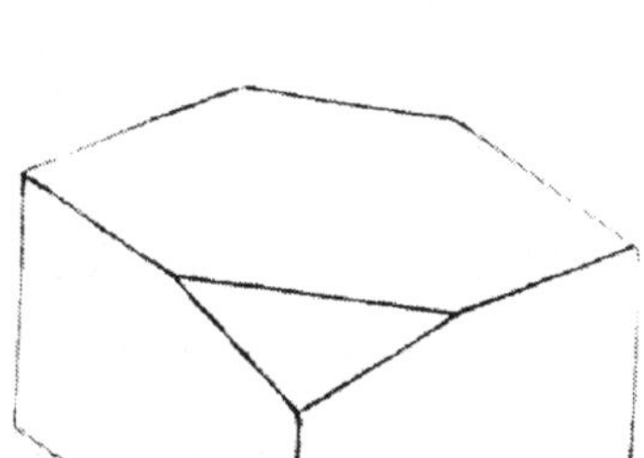

TAB. CLIII.

PLUMBUM carbonatum primitivum.

Primitive Crystallized Carbonate of Lead.

Class 3. Metals. *Order* 1. Homogeneous.
Gen. 15. Lead. *Spec.* 4. Carbonate of Lead.
Div. 1. Crystallized.

THIS specimen is a great curiosity, as it does not seem to have been before known that Carbonate of Lead crystallized in rhomboidal prisms. Haüy seemed to think the octaëdron to be the primitive figure; indeed we have been able in some specimens to trace all but four of his fractures; this however appears to be preferable, as it bears some analogy to Sulphate of Barytes, which has a rhomboidal primitive only differing in the angles, otherwise often producing similar modifications, which vary according to the nature of the substance. The angles of this rhomboidal prism are 76° and 104°; the laminæ are very distinct on all the faces. Carbonate of Lead, when we find it so nearly resembling this substance, may soon be discovered by the help of the blowpipe in procuring a little globule of Lead from it on Charcoal. They both form a nearly opaque glass; but if the heat is continued, the one will of course be reduced to Lead, and the other will remain unaltered.

TAB. CLIV.

PLUMBUM carbonatum octaëdrum.
Octaëdral Carbonate of Lead.

We are indebted for the present specimen to James Brodie, Esq., who brought it from Stolfield near Lossie Mouth, Elginshire, from an estate of John Brander, Esq. It is curious for having an octaëdral crystal nearly resembling, at first sight, the secondary one figured by Haüy, *pl.* 67. *fig.* 46. There are the four faces of the primitive octaëdron as mentioned by Haüy, *fig.* 45. M: See *the trapezoidal faces.* It has also four triangular faces agreeing with *y* of *fig.* 50.—*Haüy.*

This specimen is an example of a hard flinty rock holding Galæna or common Lead ore, sufficiently good to tempt the miner; but the hardness of the rock is an obstacle not easily overcome without an amazing expense. In these improved times, however, I should think this difficulty less considerable than formerly, if the ore is sufficiently abundant. There is perhaps an additional hope to the owner that he may not be aware of, which is, that Phosphate and Carbonate of Lead accompany the Galæna; and where these are, the rock is fissile, and more easily accessible. The situation, as to distance of fuel and conveniences for smelting, is certainly to be considered.

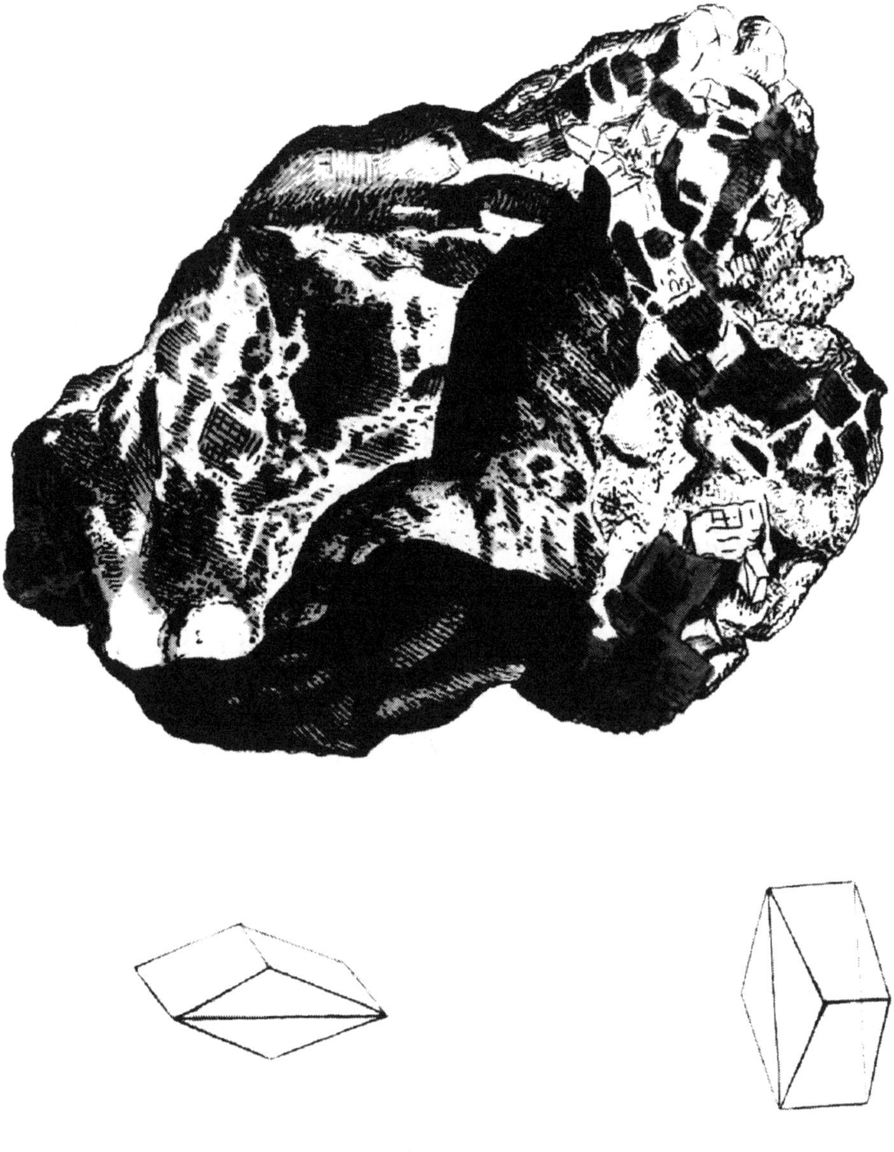

TAB. CLV.

PLUMBUM carbonatum.

Straw-like Carbonate of Lead.

Div. 2. Imitative.

SYN. Plomb carbonaté aciculaire. *Haüy*, 3. 483.

CARBONATE of Lead in spiculæ is found in some parts of Cornwall, Devonshire, Somersetshire, the county of Durham and other places where Lead ores are found, but the richest specimens are found at Wanlockhead in Scotland. The present specimen is an extremely irregular bundle of fibrous crystals, many of them nearly tubular, curiously formed among ochre which gives it an odd appearance. The spiculæ are chiefly formed into irregular columns, something like short pieces of straw. It is upon an Argillaceous Iron Stone with thin veins of Quartz.

Mr. Laing, from whom we had this, has a specimen with similar crystals seven inches in length, five in breadth, and four in thickness. The crystals are in similar groups, but cemented by amorphous Carbonate of Lead. The entire specimen is of a pure silky white colour. Large as this specimen is, the crystals are not greater than in the one figured.

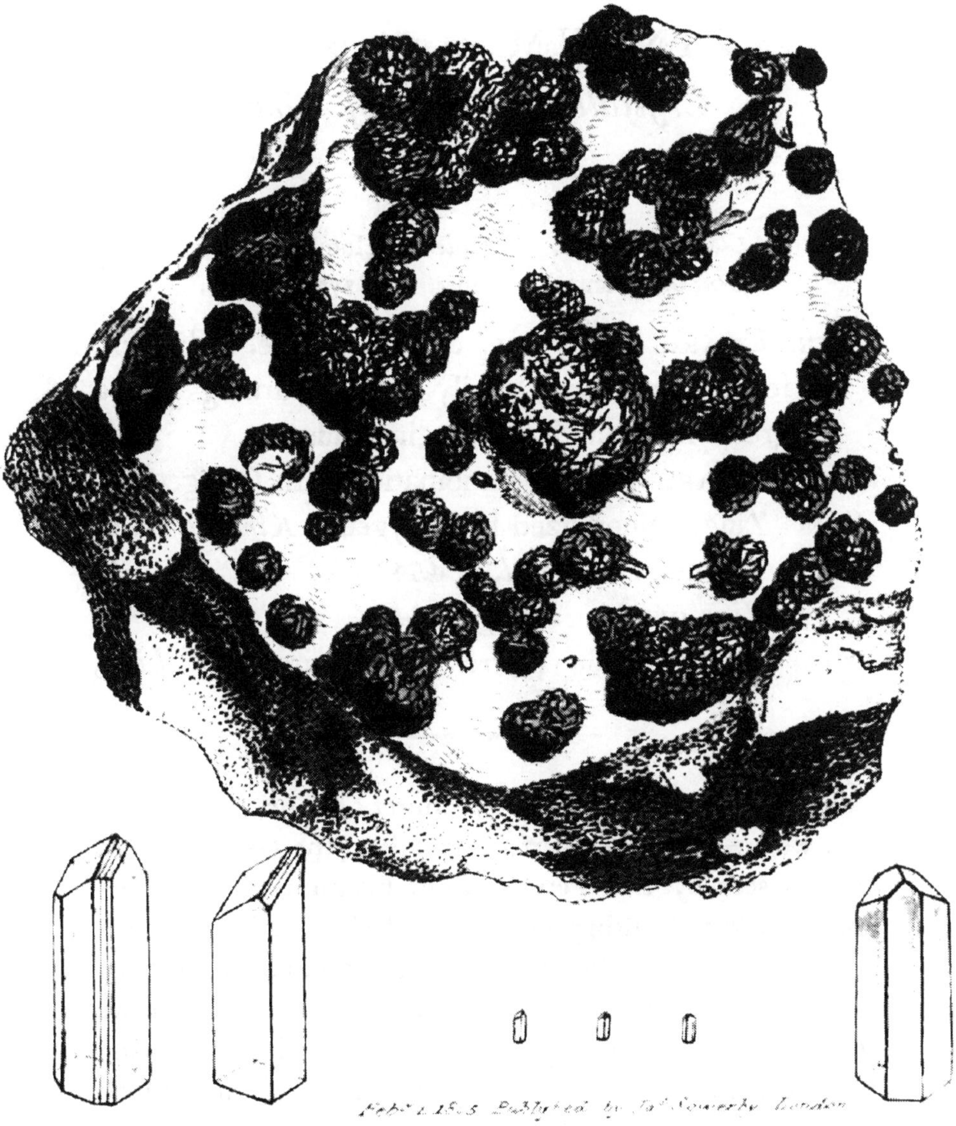
Feby 1 1805 Published by Jas Sowerby London

TAB. CLVI.

ZINCUM oxygenizatum.

Crystallized Oxide of Zinc.

Class 3. Metals. *Order* 1. Homogeneous.
Gen. 6. Zincum. *Spec.* 1. Oxygenizatum.

GEN. CHAR. Light gray, ductile, brittle. *Fracture* foliated, brilliant; easily fusible, burning with a green flame and soluble in acids.

SPEC. CHAR. Zinc in combination with Oxygen.

SYN. Zinc, mineralized by Oxygen. *Kirw.* 2. 233.
Galmei. *Emmerl.* 2. 454.
Zinc oxydé. *Haüy*, 4. 159.
Calamine, ou Pierre Calaminaire. *De Lisle*, 3. 79.

HAVING figured Blende or Sulphuret of Zinc, *tab*. 74 and 75, we are glad to add perfect crystals of the *Oxide* of *Zinc*. It was sent by Earl Ferrers, from his mines in Leicestershire, to Miss Codrington, and enriched that Lady's collection of minerals, who discovered these beautiful crystals. It is the only specimen that has yet been noticed; and we cannot help mentioning that lady's generosity in allowing us to examine it with every freedom. I thought it of much use to figure and describe it as well as was practicable from the little that can be spared from the specimen, which altogether does not contain more than a grain or two at the most. It consists of beautiful topaze-coloured crystals to be seen without the help of a lens: others, however, may be seen without one, three in particular, as figured. Hitherto crystals of Oxide of Zinc have been but little noticed in Great Britain. Mr. Smithson, in *Phil. Trans.* for the year 1803, part i. 17. after speaking of a yellowish Calamine from Derbyshire not electric, says of Electric Calamine—"that the Abbé Haüy has considered this kind as differing from the other Calamines only in the circumstance of being in distinct crystals; but it has already appeared, in the instance of the Derbyshire Calamine, that all the crystals are not electric by heat, and hence, that it is not merely to its being in this state that this species owes the above quality. And the following experiments on some crystals of electric Calamine from Regbania in Hungary, can leave no doubt of its being a combination of Calx of Zinc with Quartz; since the quantity of Quartz obtained, and the perfect regularity and transparency of these crystals, make it impossible to suppose it a foreign admixture of them. They were not scratched by a pin; a knife marked them.

"According to Pelletier's* experiments on the Calamine of Fribourg in Brisgaw, which is undoubtedly of this species, its composition is:

Quartz . . .	0·50
Calx of Zinc .	0·38
Water . . .	0·12
	1·00

and according to his own experiments:

Quartz . . .	0·250
Calx of Zinc .	0·683
Water . . .	0·044
Loss	0·023
	1·000

"The water, he observes, is most probably not essential; and in that case, from his experiments, it would be:

Quartz . . .	0·261
Oxide of Zinc .	0.739
	1·000

"He also says he has found this species of Calamine among the productions of Derbyshire in small brown crystals, &c.; and their form seems, as far as minuteness and compression together will allow of judging, nearly or quite the same as those of Regbania; and the least atom of them on being heated immediately evinces their nature by the strong electricity it acquires. On their solution in acids they leave Quartz."

When we first looked at this specimen, not having seen such before, we did not know what it was; but on trial with the blowpipe, as it passed into flocculi and dissipated, we soon determined it; nor did we consult any author, until we had made sketches of the crystals, when we were not a little pleased to find our observations agree with those of others as to the form of them. There are some additional facets not mentioned in Haüy, which, although very minute, were sharp enough to be discerned. It is very curious to see a small portion gently warmed, how readily it attracts a few scrapings of paper which cling to it for some time.

These little elegant crystals stand on a gangue of red Sulphate of Barytes crystallized on the surface in little plates of a pinkish colour, modified something like *tab.* 96, *bottom figure.* There are some blotches of marle mixed with the gangue. We also find, on the surface, Sulphuret of Zinc and Sulphuret of Copper in somewhat irregular groups of crystals; the first of a gray colour, the latter of a golden hue, and almost in tetraëdrons. There are some nearly metastatic crystals of Carbonate of Lime, and a few

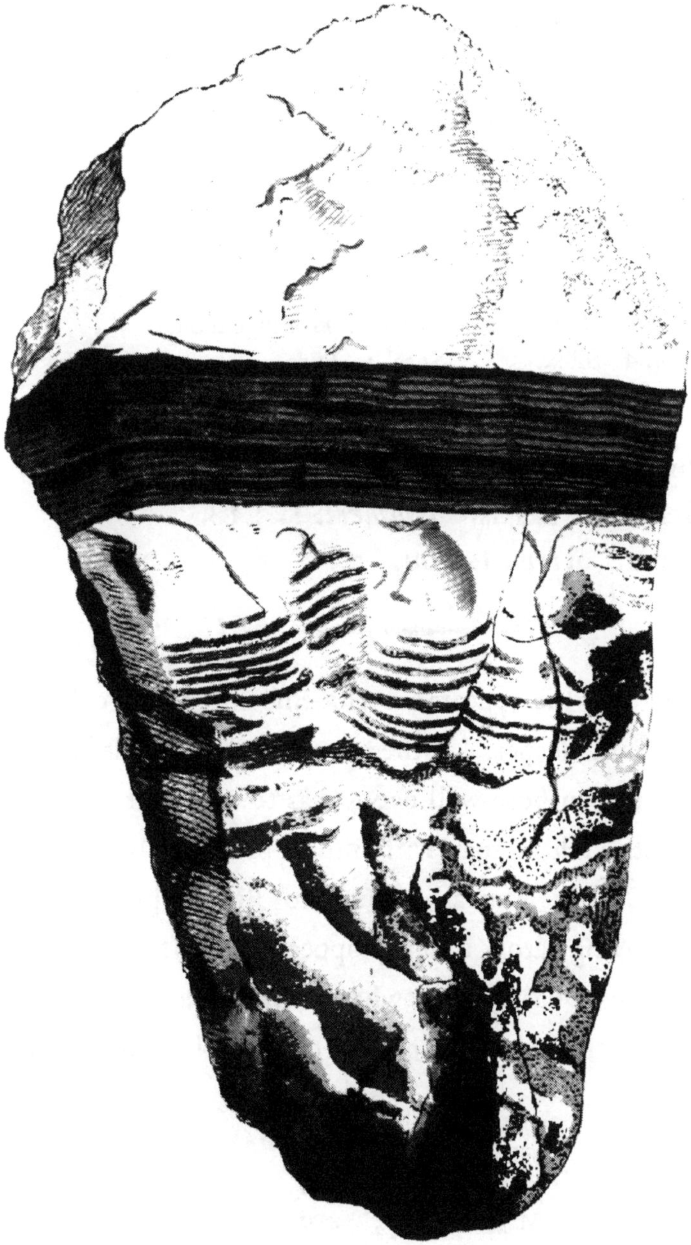

TAB. CLVII.

SILEX Quartzum; *var.* Jaspis.

Quartz-Jasper.

Class 2. Earths. *Order* 1. Homogeneous.
Gen. 4. Silex. *Spec.* 1. Pure.
Div. 3. Amorphous.

SYN. Quartz-Jaspe. *Haüy*, 2. 435.
Gemeiner Jaspis. *Emmerl.* 1. 243.
Jasper. *Kirw.* 1. 309.

SOME Jasper has the appearance of a ribband, and is called Ribband or Band Jasper. Either term is intelligible, and may answer the purpose well enough as to this part of the character, although it may with much propriety be called Stratified. Jasper is nearly allied to flint approaching hornstone, having rather a horny appearance. It is an impure amorphous Quartz, somewhat altered by a mixture of Argilla *. Its fracture is smooth, conchoidal or bluntish,

* Porcelain Jasper is said to contain:

Silica	60·75
Argilla . . .	27·25
Magnesia . .	3·00
Oxide of Iron .	2·50
Potash . . .	3·66

splintery, very little translucent at the edges. It is rather tougher but scarcely harder than flint. It occurs in many varieties, and is often marked with darker and lighter stripes, but seldom very bright. Dull green is perhaps most frequent. The present figure exhibits a very distinct neatly striped vein in part of a variegated rock composed of Quartz, &c., and there are small threads in little veins passing irregularly from it. It was picked up on the coast of Airshire, and is in the possession of my friend Mr. Laing of Edinburgh.

I have pieces of large masses of Jasper striped or coloured nearly in the same way, from the shores of Scotland. Jasper was formerly much used for large trinkets, &c., as it takes a good polish.

158

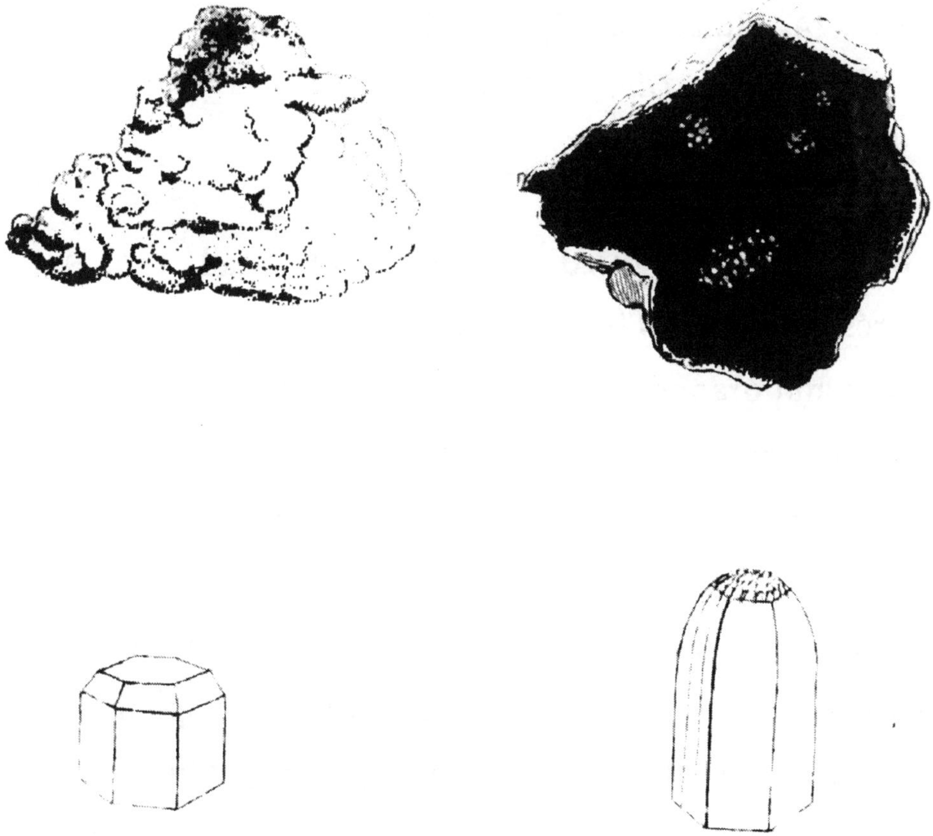

London

TAB. CLVIII.

PLUMBUM phosphatum.

Phosphate of Lead.

Class 3. Metals. *Order* 1. Homogeneous.
Gen. 14. Plumbum. *Spec.* 2. Phosphatum.
Div. 1. Crystallized.

We have already figured Phosphate of Lead, *tab.* 84. of the present work. These beautiful little specimens show the yellow and orange varieties, and two different modifications of the crystals, one bevilled on the horizontal edges of the column (see *the right hand figure*), the other rounded on the same edges, approaching such as form spiculæ, and is as it were formed of spiculæ, side by side, by which we see an example of crystals forming of spiculæ or lengthening in their modifications by spiculæ; which often happens, according to circumstances, as other substances do in plates, as has been most commonly observed. These specimens came from Wanlockhead by favour of G. Laing, Esq.

TAB. CLIX.

CALX Carbonata.

Crystallized Carbonate of Lime.

Class 2. Earths.	*Order* 1. Homogeneous.
Gen. 3. Lime.	*Spec.* 5. Carbonate of Lime.

THIS curious specimen is from Wanlockhead in Scotland. Besides the whole group being formed like a cone, the individuals are a curious crystallization with a three-sided column somewhat rounding and truncated at the apex, and a central depression. The sides of the column are again divided into four facets, better understood by the figure than by description. The apex is formed of a six-sided facet.

We thought it somewhat resembled the hard Carbonate of Lime; but the fracture and its softness give evident proofs of the contrary. We do not know that this modification is to be found in any cabinet but in that of Mr. Laing, which contains the specimen here figured. It is formed as it were from the edges of the nucleus or primitive rhomb, with the obtuse angle upwards; the laminæ decrease less than in the metastatic, and consequently form a longer pyramid, which, as it is inclined to be rounding, cannot be measured. The truncations nearly on the column and those on the pyramid give it an odd appearance, more especially that on the apex, which has a rising margin. The order of formation from regular nuclei shows how much variety nature is capable of producing by the most simple means.

The specimen is a little coloured towards the bottom with ochraceous Oxide of Iron, and in some parts are a few blotches of decomposing Pyrites.

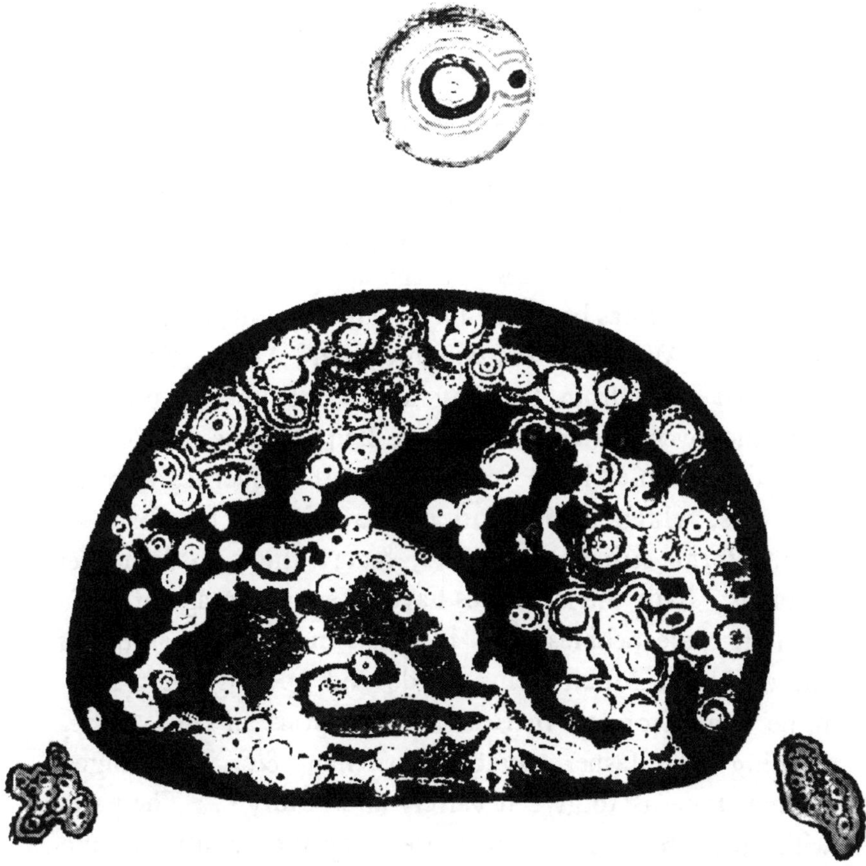

TAB. CLX.

SILEX Quartzum; *var.*

Quartz, or Agate.

Class 2. Earths. *Order* 1. Homogeneous.
Gen. 4. Silex. *Spec.* 1. Quartzum.

Div. 2. Imitative.

The river Tay in Perthshire, as well as other parts of Scotland, affords many beautiful pebbles, with circles and centres with more or less of the appearance of eyes, &c. *The upper one* has been cut for ornament, and was a present from Lady Wilson to her daughter Lady Arden. *The lower specimen* came from the Tay above mentioned, and was lent me by my friend G. Laing, Esq. These pebbles appear to be formed in the rocks in short stalactites forming in coats, which being more or less coloured by Oxide of Iron, form stripes or circles, &c.; the outside rather knobby. This the lapidaries have taken advantage of, as is seen in these pebbles, and it assists in the disposition of the colours. Much seems to depend upon Quartz, Alumine and Iron to increase the beauties of this stone, forming in it Jasper, Agate, Cornelian, &c., so arranged in one stone as to give it variety and beauty *. The parts with greenish specks, which are magnified a little at the bottom, are more porous than the rest, apparently filled with Chlorite.

* The red circles are often specks of red in transparent Quartz.

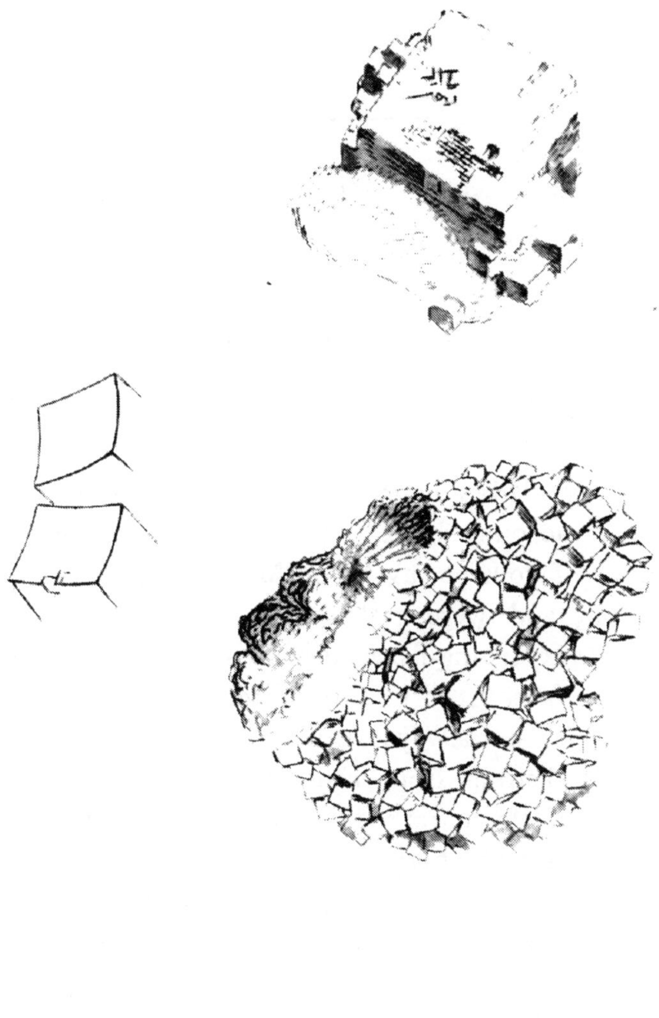

TAB. CLXI.

SILEX analcimus primitivus.
Cubical Analcime.

Class 2. Earth. *Order* 1. Homogeneous.
Gen. 4. Silex. *Spec.* Analcime.
Div. 1. Crystallized.

ALTHOUGH the Cubic Analcime of Haüy, or what is still by some called Cubic Zeolite, is not rare in some parts of Ireland, especially among basaltic rocks, yet we have had very little account of it. The present specimens were sent me by favour of my friend Mr. Templeton, F. L. S., of Belfast, and were gathered at Cave Hill, near that place. Haüy distinguishes this fossil as having fractures on the six faces*, or in three directions only; but these specimens are not always easily procured. It is, however, sufficiently distinct from Fluor by its superior hardness, easy fusion into a transparent white glass without ebullition, and want of phosphorescence; having indeed all the characters of *tab.* 59, except as to form. I have met with impressions of this with other Analcime, as well as small crystals imbedded in the radiated Analcime, or what is by some called radiated Zeolite, in the holes of Basaltic Trap, if I may so call it, for it certainly is not Lava, with which this sort of Trap has been confounded. I hope soon

* Those of the cube.

to find an opportunity of explaining the nature of Basalt and Volcanic Trap, which has caused so much argument amongst men of science as to its nature and origin.

The upper specimen shows an almost independent largish crystal.

The lower specimen is a considerable group of small crystals, with sometimes curved or concave faces, marked with diagonal striæ:—see *the lateral magnified figure.*

This substance in these shapes occurs in many parts abroad, as well as in Scotland; but I do not know that it has been mentioned as found in Ireland till now.

TAB. CLXII.

FERRUM sulphureum; *var.*
Hair-like Pyrites, or Sulphuret of Iron.

Class 3. Metals.	*Order* 1. Homogeneous.
Gen. 8. Iron.	*Spec.* 5. Sulphuret.

Div. 2. Imitative.

THIS beautiful specimen was found many years ago in the magnificent Lime Quarries of Lord Elgin, at Broomhall in Fifeshire, and is the only one of the kind which has been discovered.

The specimen is in the possession of my kind friend Robert Ferguson, Esq., who was so good as to send it from Scotland for me to delineate. The appearance of so many radiating, straight, bristle-like forms has a remarkable effect, and is very striking. Although they are extremely attenuated, they seem to be square, and are probably elongated octaëdrons. Their passing through and among the Pearlspar* so uninterruptedly is worthy of notice, and will give us reason to believe that both substances settled from their solvents at the same time, or nearly so, without disturbing each other. They are both upon shelly grayish Lime Stone. It is with much pleasure that I introduce this as a British specimen, as it is not at all generally known to be found any where but at Joachimsthal in Bohemia, and even there very seldom.

TAB. CLXIII.

SILEX quartzum; *var.* ligniformis.
Wood-like Quartz, or Petrified Wood.

Class 2. Earths. *Order* 1. Homogeneous.
Gen. 4. Silex. *Spec.* 1. Quartzum.
Div. 2. Imitative.

PETRIFIED Wood is by no means uncommon. It is, however, equally curious with many rarities, in showing some of the phænomena of the creation. Thus we find, when one substance passes away, another takes its place. This is an excellent evidence *that when one substance passes to decay it forms and organizes others.* The water that once assisted in the rise and flourishing verdure of the wood when in youth and health, now, in decay, helps to introduce a new substance in its place, and with such nicety that it is quite a deception. The colour and external appearance are still the same, although turned to stone; and the parts of the Wood, Hydrogen and Carbonic Acid are now evaporated.

This specimen was brought from Fonthill in Wiltshire by A. B. Lambert, Esq., V. P. L. S. It was eighteen inches in length, and twelve inches in circumference, and is one of the best specimens I have seen to show the nature of the change of place, viz. the Silex replacing the carbonaceous principle of the wood; the Silex in solution, as it were, taking place of the former substance particle by parso perfectly arranged and coloured, with so little disturbance, that the very cracks and broken parts are detected with the utmost precision. Possibly the Oxide of Iron, or colouring substance, does not evaporate with the other principles: thus the colour of the parts is identically preserved. The specimen is externally somewhat granular, with the appearance of a fine-grained compact Sandstone, more dense in the centre, resembling Flint, and in some parts almost Opal *.

I have received specimens of this nature from Ireland, through the kindness of Dr. Scott and Mr. Templeton, and from Warwickshire by favour of Lady Aylesford, variously stained, and otherwise acted upon in the same piece, showing that the Wood had been more or less decomposed, or was decomposing in different parts, before the metamorphosis had taken place. Since the above was written His Grace the Duke of Bedford has favoured me with a curious and interesting piece of Petrified Wood, from Aspley, which has the hollows left by some insect very perfect. A most remarkable specimen of this kind, is described in Dr. Smith's *Tour on the Continent*, v. 3. 113.

* Wood wholly opalized is sometimes found.

TAB. CLXIV.

The former specimen is externally very much allied to a Sandstone, although somewhat more condensed in the middle or centre; which often happens. *The upper speci-*

repairing the sewer there, and was given me by Thomas Furley Forster, Esq., F.L.S. It was very loose in the outer texture, and is quite solid within, losing the appearance of Wood, having the hardness and fracture of Flint, with a colour like that of Wood. It is stained black in some parts, having the appearance of burnt Wood. In this specimen we can scarcely determine whether it had been scorched or blackened by artificial fire, (as is sometimes done with Wood to give durability, and to prevent its rotting,) or by a natural process, more gentle. *The piece figured below* came from Derbyshire by favour of Mr. Martin, and is nearly black all over, excepting the outside, where it seems the bark may have covered it. This blackness gives it the appearance of having been charred by fire; but fire in the usual way must have affected the outside by stains or smoke, &c.: this therefore is an extraordinary appearance, and difficult to be accounted for with certainty. It is however to be seen, in some specimens which I possess, that the charring process may be effected by the natural progress or decomposition, which is continually seen to take place as far as our limits extend, and all over the surface of the globe. The Silex in this specimen, by attempting to crystallize, has rendered it somewhat granular, and has in some measure destroyed the finer and more delicate parts of the Wood.

I do not at present hazard a conjecture about what kinds of Woods these are. I have had many opinions from good judges about them; but must wait for further experience and more varieties. The remains of petrified woods, impressions of plants, &c. in different states are very universal. In coal-countries, and sometimes in other places, the Carbon and Bitumen often pass into Coal, or new combinations under the influence of their particular situations.

TAB. CLXV.

FERRUM Scheelatum.

Scheelate of Iron, or Wolfram.

Syn. Wolfram. *De Lisle, t.* ii. 311. *and* iii. 262.
Scheelin ferruginé. *Haüy.*

This substance had long retained the German name of *Wolfram,* although mineralogists were much divided about what class to refer it to, until Scheele had discovered Tungsten. It is found in tolerable quantity in Cornwall, and is said to be mostly found in Tin-Countries. We, however, have specimens from the Isle of Man by favour of His Grace the Duke of Athol, and Lord James Murray. I understand that Tin has formerly been known there. Much Spathose Iron Ore has been found there, very much of the same nature and kind as that figured in *tab.* 53. of this work.

The present specimen was the gift of my good friend P. Rashleigh, Esq., often mentioned in this work, and is of the more consequence as it is in some parts crystallized, so as to allow me to describe the form of its crystals, and in which state it is rarely seen.

The primitive crystal is a cube which may be fractured parallel to one of its faces, commonly with great ease, and parallel to another with some difficulty; but in the direction of the third with much irregularity. The crystals on *the upper specimen* are so small that they cannot be undersented by an outline on *the left hand of the plate* to show its modification, and another more complicated on *the right.* By examining these it will be found that they modify principally on one side of the crystal, leaving the other sometimes unchanged. *The right hand one* is more compound than any of Haüy's crystals.

The lower specimen has part of a pyramidal face exposed, and the plated fracture is very distinct as well as the shining lustre of the surface: this somewhat resembles the specimen from the Isle of Man, and is the most usual form of large ones.

TAB. CLXVI.

THE rarity of this substance in determined crystals, especially in Great Britain, is a sufficient excuse for figuring a specimen of the present appearance, which has no pretension at first sight to value. Experience, however, has shown us that in some things that rarely crystallize a characteristic specimen is a tolerable prize. The present has some small but determined crystals when examined carefully with a lens, and *the little middle figure* is a singular formation of hexangular plates, which being deposited at the lower part regularly in equal-sized plates, form an hexaëdral column. Those immediately above the column, diminishing by degrees, form part of a pyramid; and a few plates of a little larger dimensions, hanging slightly over at the apex, seem to explain by this small irregularity, that the plates formed before they were deposited. *The lower*

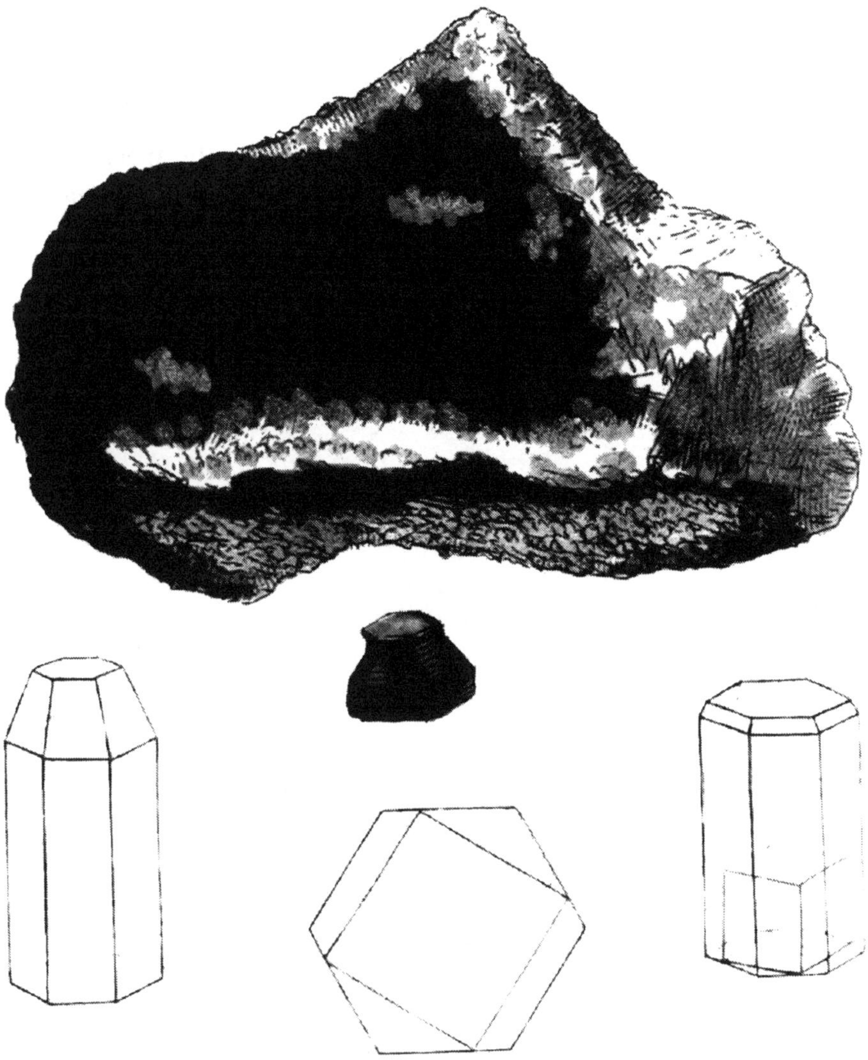

the accumulation on the sides, forming the hexaëdral plates. *The right hand geometrical figure* shows the rectangular primitive or cubic form within the column, which by a little thought may by a tyro be comprehended as the primitive form that accumulates to that of *the left hand figure*. The hexangular column has four angles of 121° and two of 118°. The gangue is composed of Chlorite, Quartz, Oxide of Tin, and Arsenical Iron, or Mispickel. The specimen comes from Cornwall.

Tungstate of Iron has not, that I know of, been found otherwise than crystallized, although the crystals are almost always interrupted. It may be known from most other substances by its peculiar fracture, which in the tabular crystals is perpendicular to their larger faces. It may be scraped by a knife, giving a chocolate brown powder.

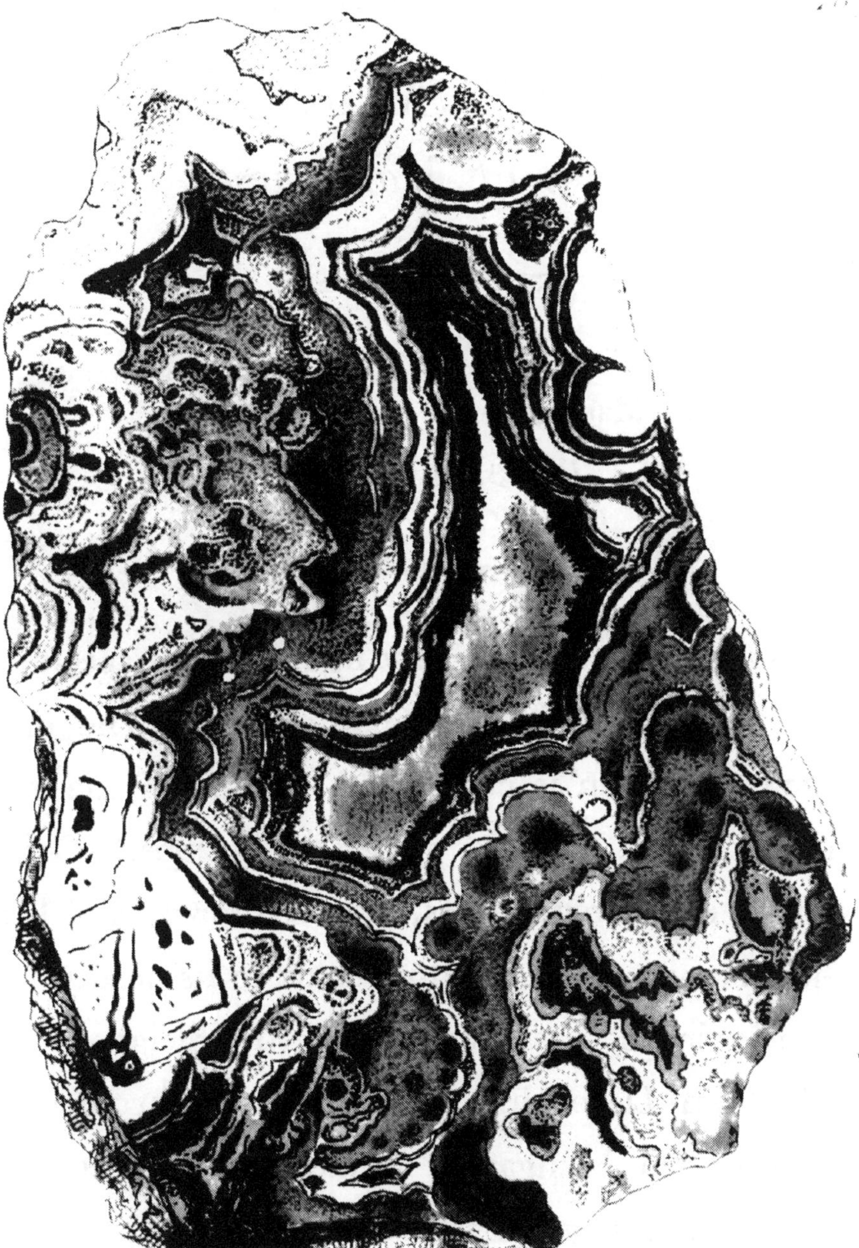

TAB. CLXVII.

MANGANESIUM oxygenizatum.

Oxide of Manganese.

Class 3. Metals. *Order* 1. Homogeneous.
Gen. 5. Manganese. *Spec.* 2. Oxide of.
Div. 3. Amorphous.

In *tab.* 86. of this work we have figured a rare specimen of Oxide of Manganese in columnar crystals. The present amorphous specimen is curious on account of the variety which it exhibits. It is supposed to be mostly Oxide of Manganese in different states, white, brown, and black; and the manner of its mixture adds to its beauty. The specimen is in the possession of Mr. Rashleigh, who received it from Devonshire. I have been favoured with specimens of red Manganese from the same place, where there is a great variety. Besides Oxide of Manganese this specimen contains in the whiter parts a tolerable proportion of Carbonate of Lime, which in some is crystallized like Pearl Spar, *tab.* 19.

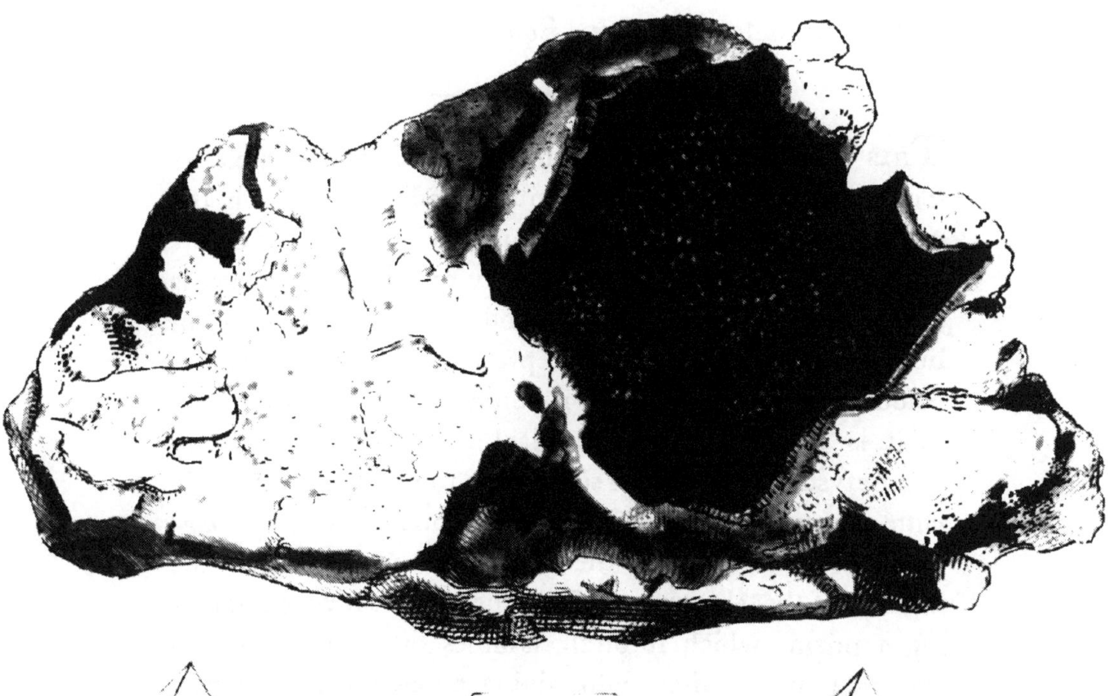

TAB. CLXVIII.

CUPRUM arseniatum.

Arseniate of Copper.

Class 3. Metals. *Order* 1. Homogeneous.
Gen. 10. Copper. *Spec.* 8. Arseniate of.
Div. 1. Crystallized.

SYN. Arseniate of Copper. *Bourn. Phil. Trans.* 1801.

THIS is the third variety of Arseniate of Copper, which the Count de Bournon calls the acute octaëdron, in which the more inclined planes meet at an angle of 84°, and the others at an angle of 68°. The first variety or obtuse octaëdron is figured in *tab.* 31. and 32., and the second variety in hexaëdral plates *tab.* 37. The present is mostly of a darker colour than the others, and is commonly of a bottle-green as it is called, viz. a dark blackish green. Like many other crystallizations it is somewhat rounding, but mostly with concave faces, which are here noticed in *the geometrical figure* as well as the straighter one, and also the manner of the elongation of the octaëdron towards forming a prism, which it often assumes by being placed on one end; thus at first sight giving a new idea, and forming a four-sided prism with a diëdral summit.

By the analysis of Mr. Chenevix, this species is found to contain,

Oxide of Copper . . .	60
Arsenic Acid	39·7
	99·7

TAB. CLXIX.

In describing the crystals on this specimen I may use the words of Count Bournon, who, after speaking of the elongated octaëdron of the former table, says, "The angles of 96° are replaced by a plane which is equally inclined on the adjacent sides, and is frequently very broad. Then the tetraëdral prism is changed into a flat hexaëdral prism having two angles of 84°, and the others of 138°." The Count never saw the angles of 84° replaced. "The average specific gravity of this Arseniate of Copper taken on five pure pieces was 4·280."

TAB. CLXX.

Mr. RASHLEIGH, with his usual ardour for the promotion of science, sent me the present specimen of Arseniate of Copper. It has the appearance of cubic crystals of Arseniate of Iron on the apex or ends of the radii of Wood Copper. However, a good crystallographer may find out the real conformation; and as it is truly curious and instructive, we were pleased to have such a specimen come under our inspection, that it may be understood hereafter. The two

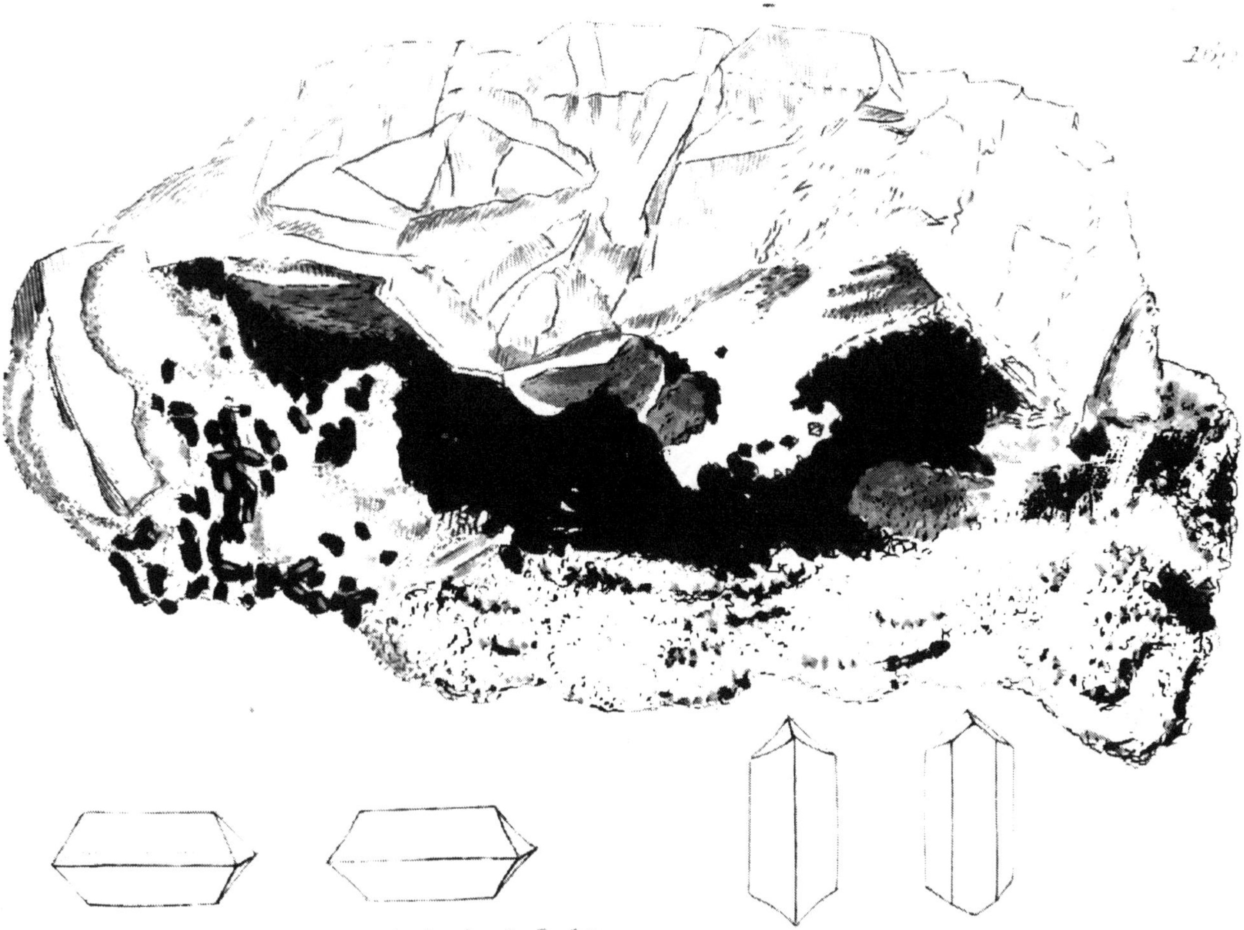

June 1 [illegible] Published by J. Sowerby London

170.

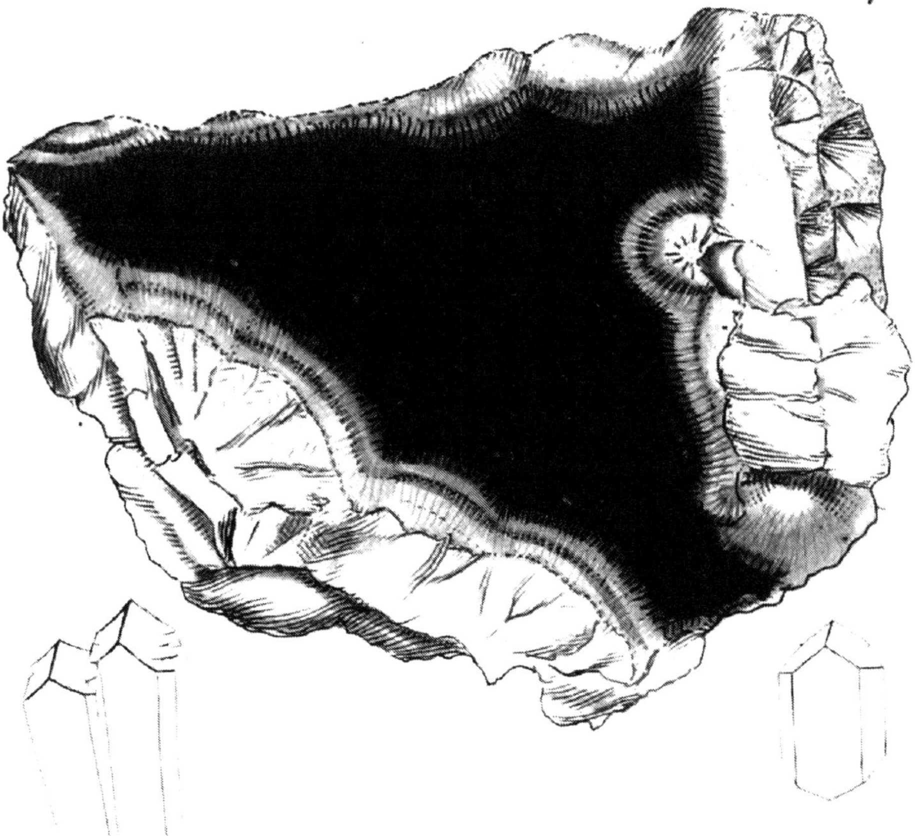

it is only a series of radiating, or partly radiating crystals, which crowd together wedge-like, as happens with calcareous Spar, terminating with octaëdrons so crowded as to show in general only as much as resembles the sides of a cube placed obliquely with the edge upwards: see *the left hand figure*. The other outlines show how it accords with the modification of the preceding figures.

There is a tendency to a concave formation of the faces which belongs to this octaëdron, and often may be seen in cubes of Arseniate of Iron, *tab.* 87 : it in some measure aids the deception. The Arseniate of Iron is generally of a yellow-green.

The Arseniate of Copper in these three tables rests upon Quartz in part crystallized, but chiefly in broken fragments mixed with Oxide of Iron.

TAB. CLXXI.

FERRUM Sulphuretum.

Sulphuret of Iron, or Iron Pyrites.

Div. 2. Imitative.

MINERALS are necessarily lessons of the changes on the terraqueous globe. The present specimen shows the cast of an *Anomia* surrounded by *Pyrites*, and the place formerly occupied by the shell remains nearly empty. It is extremely curious, that the *Pyrites*, in solution, should have formed the cast and enclosed the whole, and by some agent afterwards the shell should have been dissolved. Lord Altamont's finding *Gypsum* enclosed in *Pyrites* (and I have myself found it occasionally since) would, perhaps, account for this, if we had found *Gypsum* in the place where the shell had been, or near it; for the sulphur in an acidulous state might have combined with the lime.

We are obliged to Mr. Weeks, of Hurst Pierpoint, for this specimen, gathered by himself in that neighbourhood, curious for many interesting fossil productions.

TAB. CLXXII.

BARYTES sulphata; *var.* stellata.
Stellate Sulphate of Barytes.

Div. 2. Imitative.

Syn. White semi-pellucid Spar. *Woodward*, 88. *spec. a.* 16.
Starred waxen-vein. *Grew's Musæum, p.* 312.
Lepastrum. *Hill. p.* 146. *tab.* 2. *spec.* 1. 2.

This substance was once taken for Gypsum, or Sulphate of Lime; but is since found to be Sulphate of Barytes. To those that examine the crystallization it will be easier distinguishable, than by any other external character that we know of.

These varieties of Sulphate of Barytes are chiefly found on the western side of the island of Sheepy, in Kent; and we do not know them to have been found so perfect elsewhere in England, or in any other country. The marley cliffs of that place, perhaps raised by the deluge, and full of a great variety of antediluvian relics, have been for years falling down in small or large masses. And from these cliffs are to be seen lumps of marle from the size of an ostrich egg* to several feet in diameter; in which these Sulphates of Barytes are concealed till the masses fall or break to pieces

* Those about this size are sometimes called sea eggs by some of the guides; and when the divisions are lined with the yellower carbonate of lime, they have a more apt resemblance to eggs.

on the shore, or are broken (as they often are) on purpose for examination. The larger lumps (commonly called *Septaria*, and formerly *Ludus Helmontii* *) most generally contain them in greatest perfection among the divisions or sort of cracks in the insides.

The upper figure is prettily relieved by the delicate yellow carbonate of lime, or waxen vein, as it is commonly called, which fills up the divisions, and the Lepastri are generally fixed on the calcareous partitions, which are often crystallized, varying in colour, thickness, and number of coats. This specimen is remarkable for having a star on the side, placed immediately on the *argillaceous marle*, which is represented divided by the carbonate of lime.

TAB. CLXXIII.

THE *Lepastri* vary in general form, but not much in their crystallization; that is to say, they vary in the height, largeness, and spreading of the group. Thus there are four the most general appearances of these aggregated stellæ, which are perhaps sufficiently described by the figures. It may however be observed, that the ends of the crystals generally incline downwards; indeed they are rarely otherwise †.

The piece of *Septarium* is covered with two coats in this specimen, as well as the last, but is of the more usual colour. The coat on the *left hand side* appears to be just forming.

* Of which we shall speak more at large in another place.

† We shall explain the crystallization with some larger varieties of Sul-

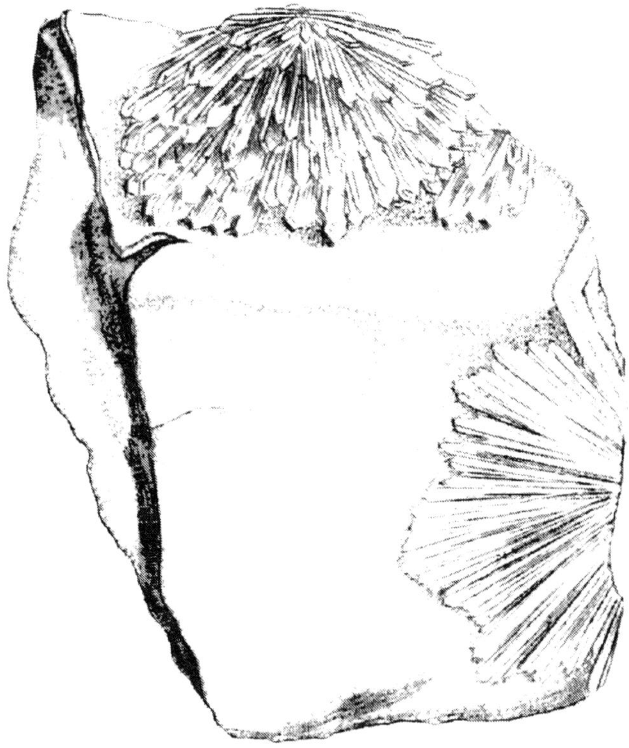

TAB. CLXXIV.

CALX carbonata; *var.* radiata.

Madreporite? or Radiated Carbonate of Lime.

Div. 2. Imitative.

SYN. Madrepor-stein. *Journal des Mines*, *n*. 47. *p*. 831.
Madreporite. *Haüy*, 4. 378.

When I gathered this specimen, which seemed to have fallen from the rock at Bow-and-Arrow Castle, on the Isle of Portland, I was instigated to keep it as of an uncommon construction, to serve as a memorandum of the place, and to recall to my memory the nature of that part of my excursion. This sort of relic, though sometimes not very handsome, is nevertheless pleasing; and it is a continual enjoyment of a pleasure to be reminded of it.

This has been rendered of more consequence, since it is a curious example of the utility of the knowledge of crystallization, which in this case distinguishes it from a coral; and might have made the substance called Madreporite better understood. It is said, "the Madreporite belonging to the class of calcareous stones, found by M. de Mollé some years ago at Russbachthall, in the county of Salzbourg, is a stone of transportation. Some specimens weigh from twenty to thirty pounds."

"Externally it resembles Basaltes, so much, that some Mineralogists considered it to be the same; others believed that it was produced from Madrepores; but it discovers no certain characters of a primitive organic formation: besides, it has so great a resemblance to the real Madrepores, that it has thence borrowed its name. It is of a gray colour, composed of divergent prisms, brilliant on their transverse fracture, and of a black and duller colour on the longitudinal fracture. The fracture exhibits a tissue of small bent laminæ; it is entirely opaque, brittle, rough to the touch, and of moderate hardness; the intervals between the bundles which compose it are in part filled with small white leaves of Calcareous Spar."

There are three or four analyses of this substance. According to Mollé, a hundred grains of it contain,

Lime..	63 $\frac{4}{10}$
Alumine	30 $\frac{2}{10}$
Iron..	10 $\frac{10}{15}$

Analysed by Klaproth, it was found to contain,

Carbonate of Lime	93 00
——— Magnesia	0 50
——— Iron	1 25
Charcoal	0 50
Sandy Silex.	4 50
An atom of Oxide of Manganese	
	99 75

Haüy's experience gave him reason to think himself satisfied that this substance, which has been made so conspicuous, is a Carbonate of Lime; and my specimen is a confirmation of this opinion, being certainly nearly the same thing; but now the improved knowledge in crystallography helps us more readily to decide it. Like the Madreporite, at first sight it looks like Basalt, and somewhat resembles a Coral, or Madrepore, and shines on the transverse fracture; viz. on the faces of the primitive rhomb which discover it: indeed, we know of nothing in the description that does not sufficiently accord with ours. The small bent laminæ are a consequence of the radiation—see *tab.* 131. This is an example of one of the dullest specimens of Carbonate of Lime of the crystallized and divergent kind, dulled with adventitious matter: thence the variations of the analysis.

175

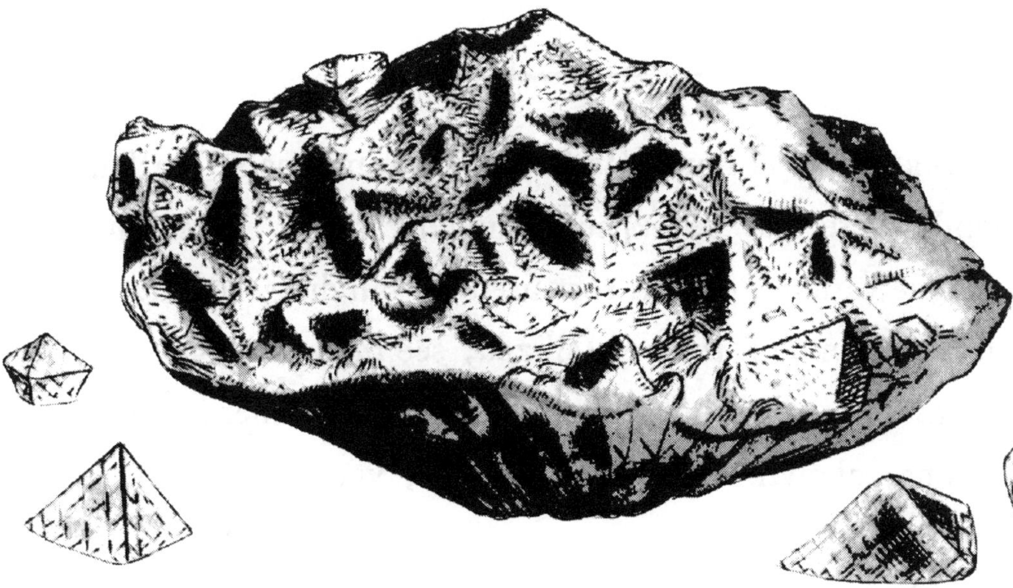

TAB. CLXXV.

CALX carbonata; *var.* inversa.
Inverted Carbonate of Lime.

Div. 2. Crystallized; *var.* inverted rhomb.

WHERE there is much Carbonate of Lime, as at Portland Island, there may be expected much variety. Thus there are some varieties of that sort used for building, see *tab.* 8; and also some of the crystallized, see *tab.* 143, 144. As well as *tab.* 174, the present specimen is rather a whimsical one, showing how nature sports, yet conformable to a certain regularity.

The last aggregation of molecules had apparently formed nuclei of inverted rhombs, arranging themselves something like an inverted pyramid—see *left hand figure, tab.* 174. The present specimen has a great variety of three-sided pyramids, nearly regular, or with prominent sides, (see *the left hand lower figure,*) and consequently concave faces, (see *the right hand figure,*) and with various kinds of bases, as in the other separate figures.

176

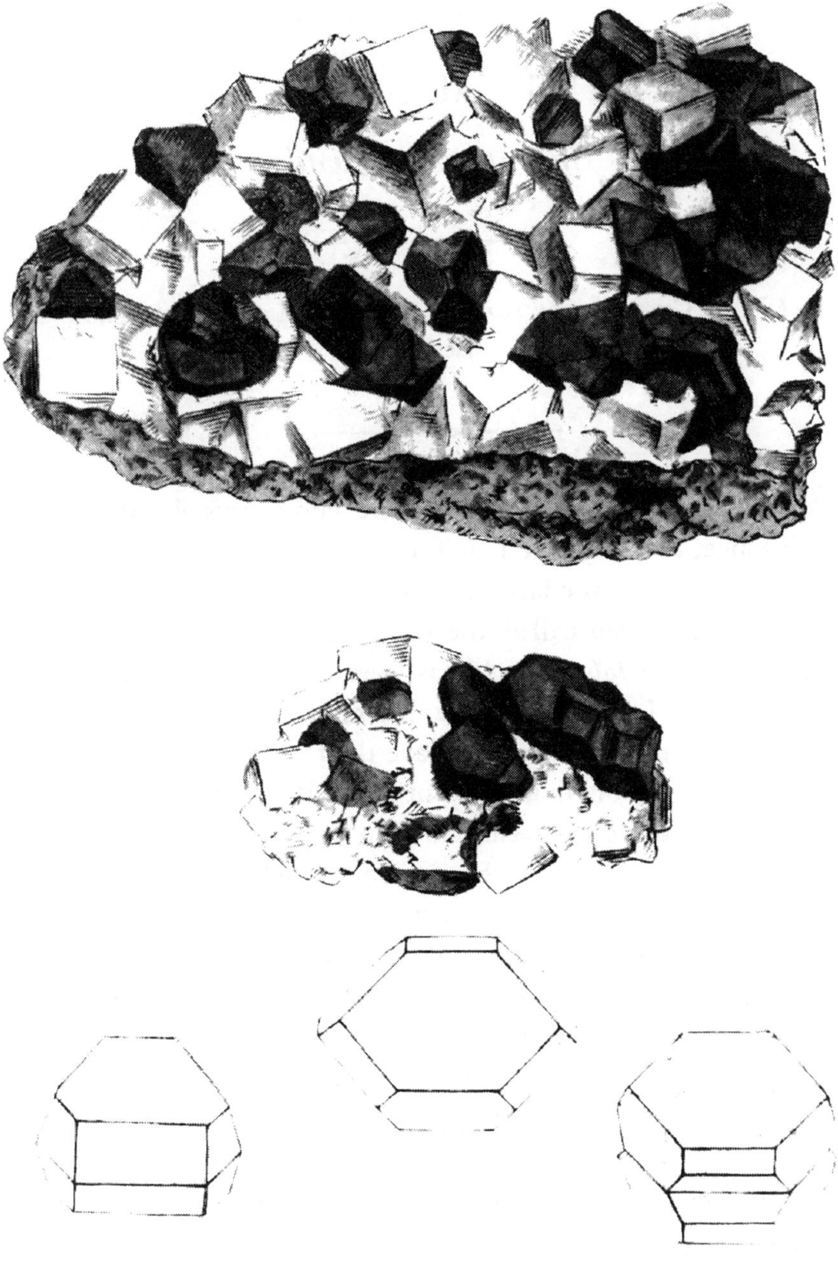

TAB. CLXXVI.

PLUMBUM sulphureum.

Sulphuret of Lead.

Div. 1. Crystallized; *var.* Cubo-octaëdron compressed and mackled.

We have exhibited the primitive cubic crystals of this substance, *tab.* 24. It is not rare to see the corners truncated, which are the faces of the octaëdron; thus it passes to what has been called the cubo-octaëdron—see the description, and *tab.* 99. It has been lately observed, that Sulphuret of Lead, or *Galæna,* has rarely been found in six-sided tables; and those who have the few so found, think much of them on that account; and it is certainly a curiosity, when we see a substance so decidedly fracturing in cubes to find it forming the octaëdron, which it often approaches, and sometimes does most perfectly—that a combination of these forms should be deposited, and at the same time with two of the faces much broader than the rest, or other twelve; viz. six hexangular faces of the octaëdron, and six rectangular faces of the cube—see *the middle geometrical figure:* that these should again be so deposited in crystallizing, as for two to meet with two of the broader faces as a basis to each, of the nature of that of the mackle, *tab.* 33; and like that they will meet at opposite angles, as if turned on an axis. Thus the six-sided faces of the octaëdral formation, and the right-angled ones of the cube formation, may meet opposite, as at *the right hand figure*, or alternate, as on *the left hand figure*.

The upper figure is from a specimen in the Marchioness of Bath's cabinet, and contains most of these modifications. *The lower one* is from a specimen in my own collection. They both came from Derbyshire some time since, and are the only British ones we know of.

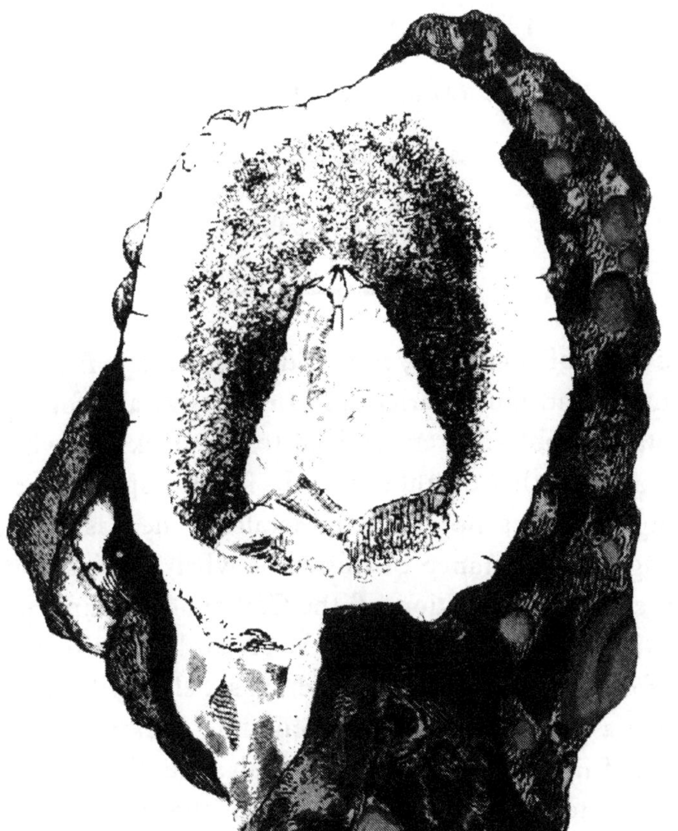

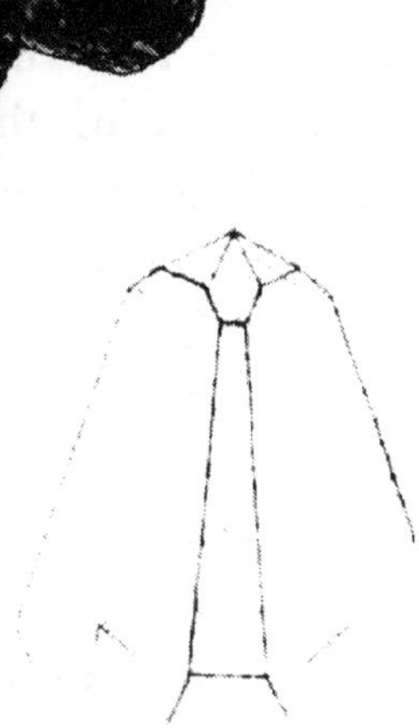

TAB. CLXXVII.

CALX carbonata.

Carbonate of Lime.

Div. 1. Crystallized.

THE hills of Pentland, near Edinburgh, are famous for Petunse*, and much variety in Mineralogy; also for Agates and other Pebbles. I am obliged to Mr. Jackson, who was botanizing in that neighbourhood, for this specimen; for, although he does not study mineralogy, he was struck by the singular appearance of this stone when he picked it up, and the regular formation of the Carbonate of Lime placed so distinctly within the hollow is certainly worthy of notice; for we know of no theory which satisfactorily accounts for such a formation. The surrounding pinky Quartz, in bundles of little eighteen-sided crystals, lines the cavity, and the jasperine Quartz (if I may so call the red coat) seems to terminate the whole pebble as it were, which is surrounded by part of the rock of a brown hue, called trap, in which there are smaller or larger pebbles sometimes included, and sometimes hollows where others have been entrapped. These hollows are sometimes coated with a green or blue earthy substance called by some the Green Earth of Verau, probably owing to an uncertain mixture of Iron: this occasionally coats also the stones included. The

* An interesting substance used in porcelain.

Carbonate of Lime is composed of half of a very acute rhomb with three largish faces of the æquiaxe, and three smaller ones, probably belonging to the primitive rhomb: see *geometrical figure.*

It is not a little singular that the like stones excluding the trap have been found in Wiltshire at a small depth under ground, of which I have obtained a specimen by favour of the Marchioness of Bath.

This specimen, rich with information, is particularly worth the attention as well of the novice as of the adept, considering the curious divisions of formation in the different substances of which it is composed. The part of the rock this came from seems to have been a mixture, as it were a chaotic one, (if I may so term it,) appearing like the fragments of various rocks that had undergone the action of moist elements, so as to form air, and water bubbles, which could not immediately escape. Apparently a continual deposition has taken place, more still forming, and enclosing the preceding till the whole matter was deposited. In the mean time each elementary substance, according to the particular formation of its molecules, and the nature of its nearest neighbour, formed, either by itself or into combination. Thus the Carbonic Acid and Lime united together, so as to construct a Crystal in the middle of this hollow as complete as circumstances would admit of, depending on the quantity of Carbonate of Lime received in solution, perfecting some faces and depositing the other molecules irregularly. A small tinge of Iron stained the solvent, and consequently the Crystal towards the top is a little coloured. The surrounding Quartz has also crystallized under similar circumstances, and is somewhat stained with the Oxide of Iron among the Crystals, giving this lining a pinkish hue, which is again conspicuous at the outer side and edge next the piece of compound rock.

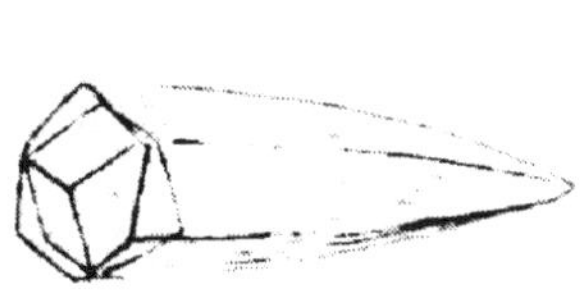

[illegible] Publ. by Jas. Sowerby London.

TAB. CLXXVIII.

CALX carbonata.

Fasciculated Carbonate of Lime.

Div. 1. Crystallized.

These Crystals of Carbonate of Lime, I believe, are rather scarce. The representation of a regular Crystal so neatly formed by a bundle of spiculæ is truly curious; and it is generally found that they have a light ochraceous or brown tint with an opaque glaucous appearance, prettily contrasted with the brilliancy of the Quartz on which they lie. This is, we understand, one of the rarest of the numerous productions of Derbyshire, and is found at a considerable depth in the ground.

179

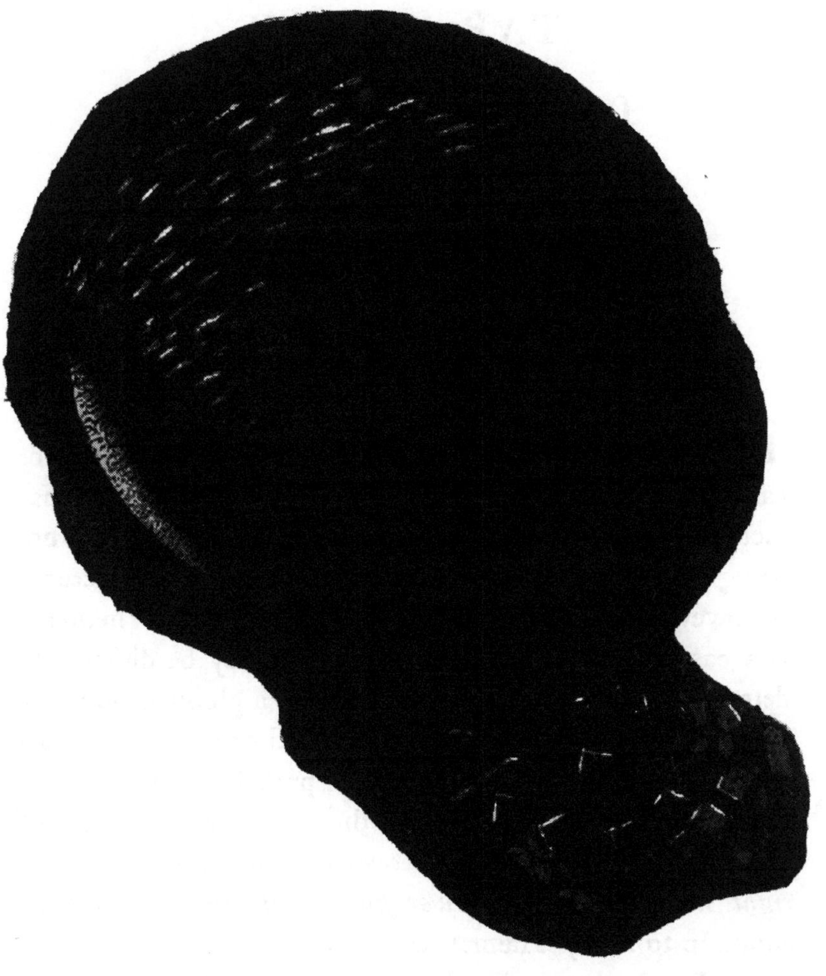

Aug 1 1806 Published by Jas Sowerby Lo

TAB. CLXXIX.

FERRUM oxygenizatum.

Oxide of Iron.

Div. 2. Imitative.

This specimen, although generally speaking it might be called an Hæmatitic Iron Ore, having most of the characters belonging to it, differs from that substance in not being striated, and in having more of an earthy fracture; it however may be placed near the Hæmatites. Whether it is a cast or change after crystallization may be difficult to determine. The lump at first looks like a globular Sulphate of Barytes (see *tab.* 96.) with the plated or tabular Crystals standing edgeways, but these are probably derived from the decomposition of Pyrites; they are however each of them so much blistered or covered with bubbles (see *the right hand side of the lower figure, tab.* 56. a formation common to many Hæmatites) that they cannot be made out. The hollow is lined with very thin-edged, probably tabular, Crystals, belonging to those of *tab.* 60. These are covered with a dark red or somewhat crimson powdery glimmer, sparkling like that of *tab.* 64. The stalk, or lower part, seems to have been cubical Iron Pyrites: see *tab.* 29 and 30, which are externally covered with bubbles. Possibly the whole once formed a Sulphuret of Iron or Pyrites, and having lost the Sulphur has become a mere Oxide of Iron. Such fortuitous circumstances as this was formed by, sometimes give the shapes of pears, apples, lemons, &c. which are as highly valued as if true petrifactions, and are very often difficult to be accounted for, especially by those who have not seen a series of specimens.

This specimen was found about the year 1799 near Staunton, about five miles from Ulverstone, Lancashire, and in that neighbourhood smaller balls have been found with the appearance of the inner part of this, which has been called velvety, downy, &c.

Published by Ja. Sowerby London.

TAB. CLXXX.

SILEX Mica.

Mica.

Gen. 4. Silex.

SYN. Mica, Muscovy Talc. *Kirwan*, *v*. 1, 210.
Mica. *Haüy*, *v*. 3. 208.
Mica membranacea. M. laminosa, &c. *Linn. ed.* 13. *t*. 3. 58.
Glimmer of the Germans.

MICA, μικκὸς, or μικρὸς, has been long noticed as a glittering substance, and often serves to countenance the idea that our streets are paved with gold and silver. It is one of the first things that takes the attention when fragments of the granite stones from Aberdeen are lying in our streets for paving or building, which has been frequent since the improvement of forming such durable stones into shape for that purpose; an improvement not much above half a century old. It soon becomes as it were familiar to the Mineralogist by its frequent occurrence; yet there is some confusion with regard to its identity, as it has been confounded with Talc: see *tab*. 182, 183, and 184.

The present specimen has most of the characters belonging to this very curious substance; it has signs of the integrant molecule, a triëdral prism, whose base is equilateral, and of the rhomboidal prism formed of two of these its primitive, with various lengthened rhomboids, hexagons,

&c.: see *figures*. It sometimes has the laminæ bent. Its lustre is of the changeable kind; *the upper plates* showing the accumulated refracted silvery opacity between those beneath, each plate being transparent if separated *. It is flexible and elastic; the plates, when bent, will return to their places with a considerable spring (see *bottom figure*, a plate bent thus much will return—the straight line seen through shows its transparency): this elasticity, and its being destitute of unctuosity, distinguish it from Talc, as hitherto the varieties of one species have been placed among those of the other by several mineralogists.

Muscovy Talc of the older authors is undoubtedly Mica, so named in contradiction to Venetian Talc; which, although nearly allied, is yet a distinct species, and may be looked on as a good sample of Talc.

TAB. CLXXXI.

THIS specimen differs very little from the Mica from Muscovy, which is occasionally found some feet in diameter. Scotland and Cornwall produce it three or four inches square, or even larger. This kind of Mica is used for lanterns, &c. and is very convenient for ships, as it bears the explosion of cannon without cracking. It is however liable to get scratched, and becomes rather opaque by heat; which opacity is not removed by water. This substance is spoken of as introduced into Rome in the time of Seneca to admit

* Its fine lustre we cannot imitate without hazard of the colour changing; we therefore must request our friends to consider the shining parts of the plate as of a fine silvery or pearly lustre.

181

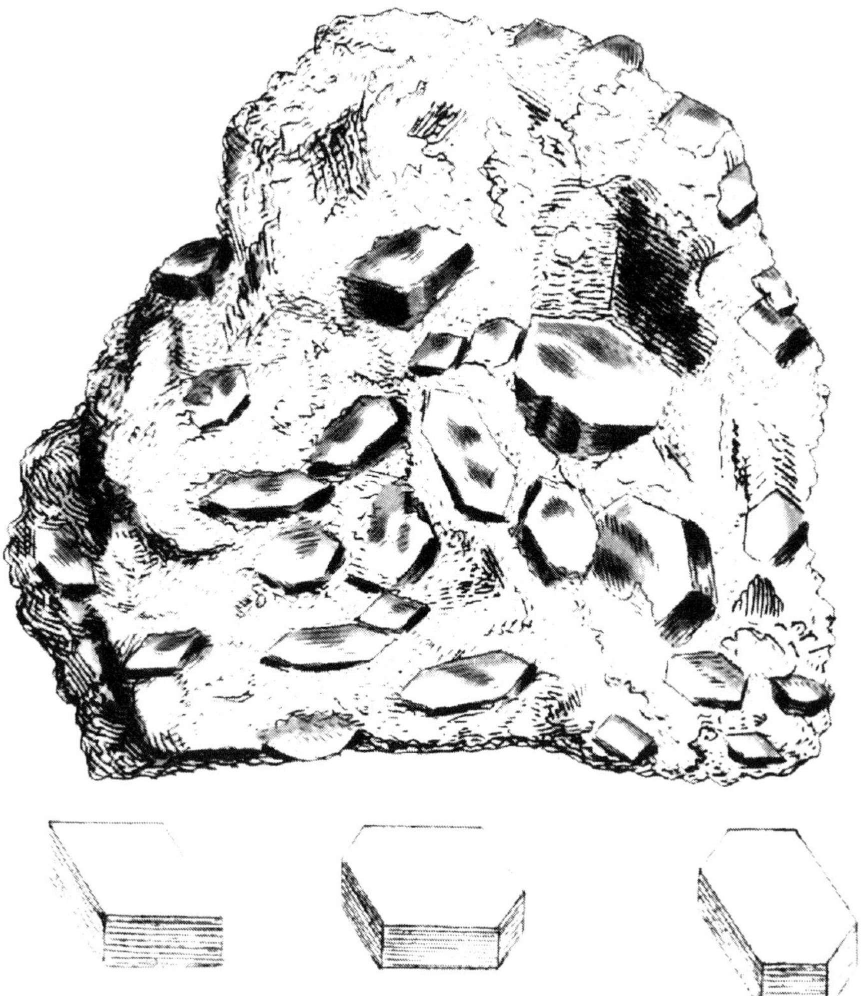

light into their apartments; but Agricola considered what was then used as a plaister-stone*, (Sulphate of Lime, *tab.* 67.) saying that although it was not affected by the heat of summer or cold of winter, yet wet wasted it considerably. These characters confirm his opinion; for we know that Sulphate of Lime is found in France in crystallized masses sufficiently large to form, when split and cut, squares of many inches diameter. The name Talc has been applied to any laminated substance.

Mica analysed by Vauquelin was found to contain:

Silica	50·0
Argilla	35.0
Oxide of Iron . ..	7·0
Magnesia	1·35
Lime	1·33
	94·68
Loss	5·32
	100·00

* It seems to have been used in Agricola's time to shelter plants; in which case it would be much exposed to weather.

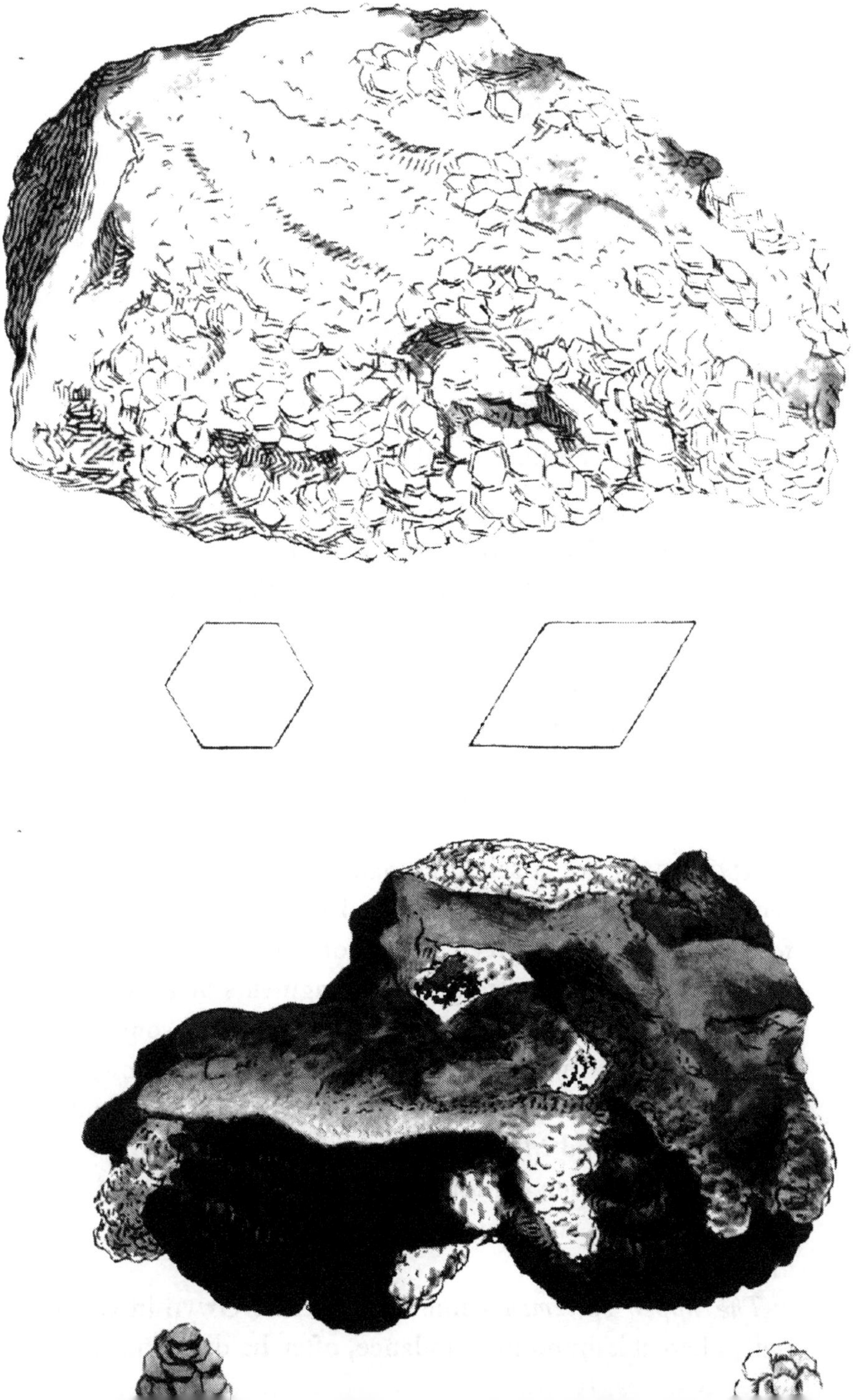

TAB. CLXXXII.

SILEX Talcum.

Talc.

Div. 1. Crystallized.

SYN. Talc, Venetian Talc. *Kirwan*, *v.* 1. 150.
Talc. *Haüy*, *t.* 3. 252.
Chlorite. *Kirwan*, *v.* 1. 147.
Talcum viridans. T. lamellare, and many others. *Linn. ed.* 13. *t.* 3. 51 *et seq.*
Mica Talcosa. *Ibid. t.* 3. 59.
Talk. *Emmerling*, *t.* 1. 391.

Chlorite is not a scientific name, according to our present improvements in mineralogical knowledge, as it is perfectly confined to a green variety of Talc, found either in small laminated crystals, granular fragments or masses, or disseminated through Quartz, &c. Its grain is sometimes very fine. I find in this latter state it is scarcely known by any one, as any green earthy substance may be confounded with it. Talc and Mica, as observed under *tab.* 180, have been till lately much confounded; but since the publication of Mr. Kirwan's excellent work they have been more and more defined.

The upper specimen came from Stenna-Gwyn in Cornwall, where it is found in abundance, often holding Phosphate of Lime, among Quartz, and decomposing Feltspar; whence the rock itself is often called Apatite, the old name for Phosphate of Lime. Abundance of Oxide of Tin often accompanies it. This is a variety of Talc, agreeing with that called Chlorite in every external character except colour, which instead of being green is that of cream.

The lower specimen has the colour most usually allotted to Chlorite, which is undoubtedly to be attributed to a large adventitious mixture of Iron, which at the same time renders it easily fusible.

Mica and Talc seem to have the same forms in their Crystals, but they are never by any means so large or well defined in Talc. These Crystals are remarkably soapy or greasy to the touch, particularly if bruised. The laminæ are very easily broken, and are apt to divide into minute greasy scales; which property has rendered this fossil an ingredient in cosmetics. The white varieties are difficultly fusible.

White Talc analysed by Hoëpfner was found to contain:

Silex	50
Magnesia	44
Argil	6
	100

Chlorite by Vauquelin:

Silica..	26·0
Argilla..	18·5
Magnesia..	8·0
Oxide of Iron	43·0
Muriate of Soda and Potash	2·0
Water	2·0
Loss	0·5
	100·0

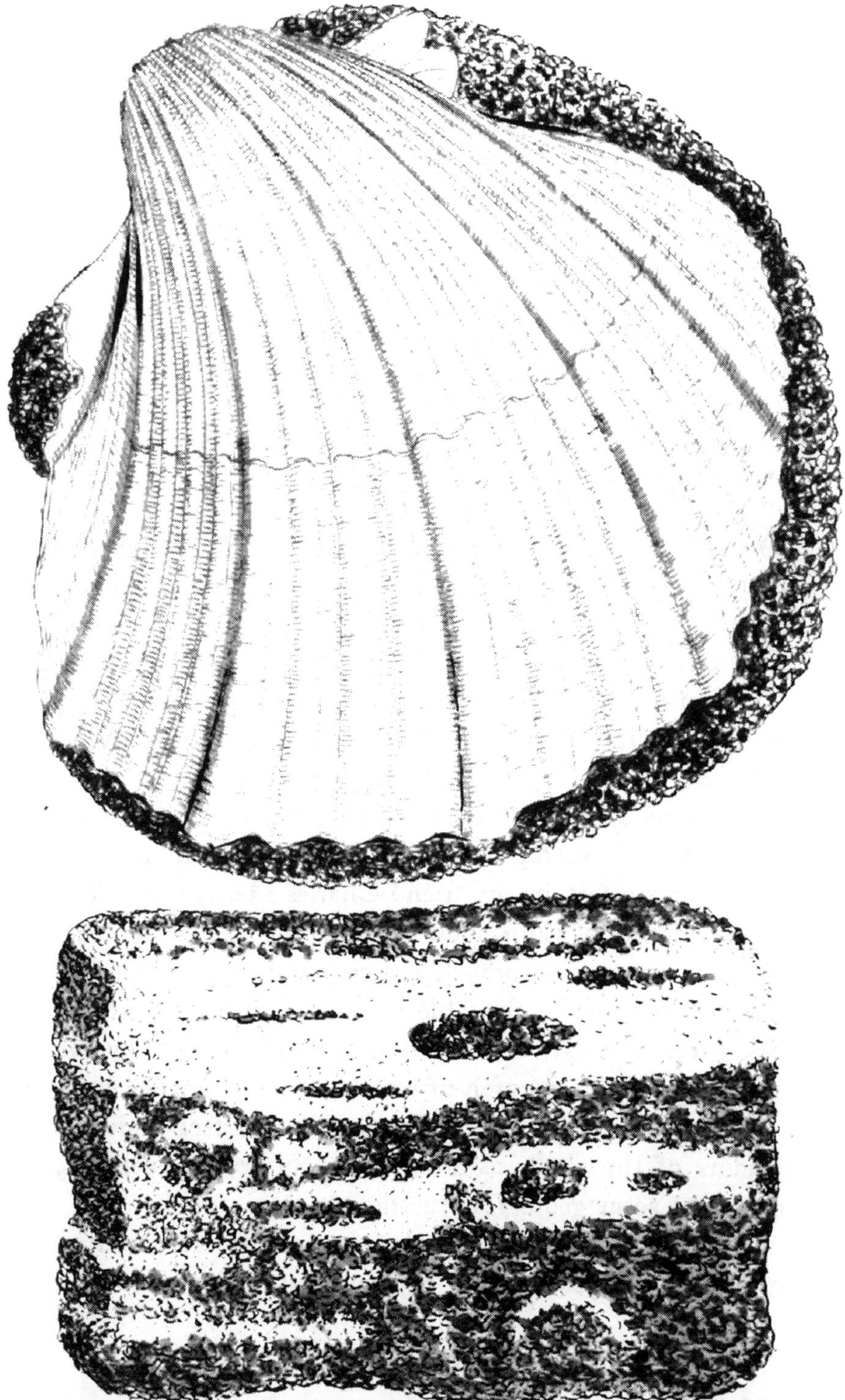

TAB. CLXXXIII.

SILEX Talcum.

Talc.

Div. 2. Imitative, in grains.

When it is found in a state as if it had suffered a change after parting with the original rock, and with the remains of other subjects and animal exuviæ, it assumes a new appearance. Thus it is found heaped in abundance in strata that seem to have originated from the fragments of rocks, as in the sandy marle * of many parts of Somersetshire, Wiltshire, &c.

The upper specimen came from Stourhead near the seat of Sir Richard Hoare, Bart., and was picked up on that gentleman's estate by my friend Charles Mead, Esq. The shell is formed of Lime, and the sand has some fragments of Lime among it, probably more or less composed of broken shells, while the Chlorite is interspersed, forming dark green specks. The shell is curiously divided by five large ridges, between each of which are three smaller ones.

We take advantage of figuring the perfect and rare remains of this shell, as it may be of future use, perhaps, in settling some geological questions.

I gathered sand nearly of the same nature at Charlton in Kent, some years since, curiously mottled or stratified

* The provincial name for this is good, being Green Sand.

with Chlorite: see *the under figure;* and Woodward mentions greenish sand from Woolwich, *p.* 11. Lord Altamont, Dr. Clarke and Mr. Warburton have since favoured me with some sandy Lime from the Castle Hill, near Cambridge, where it is very abundant, and contains numerous petrifactions.

TAB. CLXXXIV.

This forms an appearance not unlike Mortar with the green Chlorite among it. When it has a more perfect appearance of mortar without the green, it is considered as good manure, and is provincially called Goult. It is often found about a foot from the surface. This is chiefly used to make the best white bricks of Cambridgeshire. That with granular Talc or Chlorite is of a dull hue, and is found deeper. The same substance, somewhat more compact, is called in Ireland Mulatto-Stone: see *tab.* 185. I have figured some of the petrifactions that occur, considering them as useful to mineralogy. Those in this stone are generally of a dark brown stony appearance, accompanied by rugose lumps of various sizes of nearly the same substance, somewhat similar to the swampy Iron Ore of Kirwan, *v.* 2. 183. The petrifactions are coloured like it, sometimes with a nearer approach to the hue of Pyrites. These petrifactions are the round one *on the left hand,* supposed a hinder tooth of some fish; *the right hand* is considered as a fish's bony palate; *the middle upper figure* is a bivalve shell, the upper valve remaining in the state of Carbonate of Lime, the lower one browned with the ore. It is a kind of Anomia called a Gryphite, very frequent among

181

Published by Ja.s Sowerby London

petrifactions. *On the left,* under that, is part of a Coral, near which is a fish's palate, such as often extends to the form of a Lupin pod. *The left hand figure below* is part of a *Cornu Ammonis.* The rude lump of the brown ore *on the right,* with the adhering shells, like those of the *Anomia Squamula,* found sticking on crabs, oysters, &c., has the impression of the deeper valve of a small Gryphite.

In all our specimens of green *sandy* Talc, it has been so worn and rubbed about that it has lost its laminated texture, and become earthy in its fracture.

185

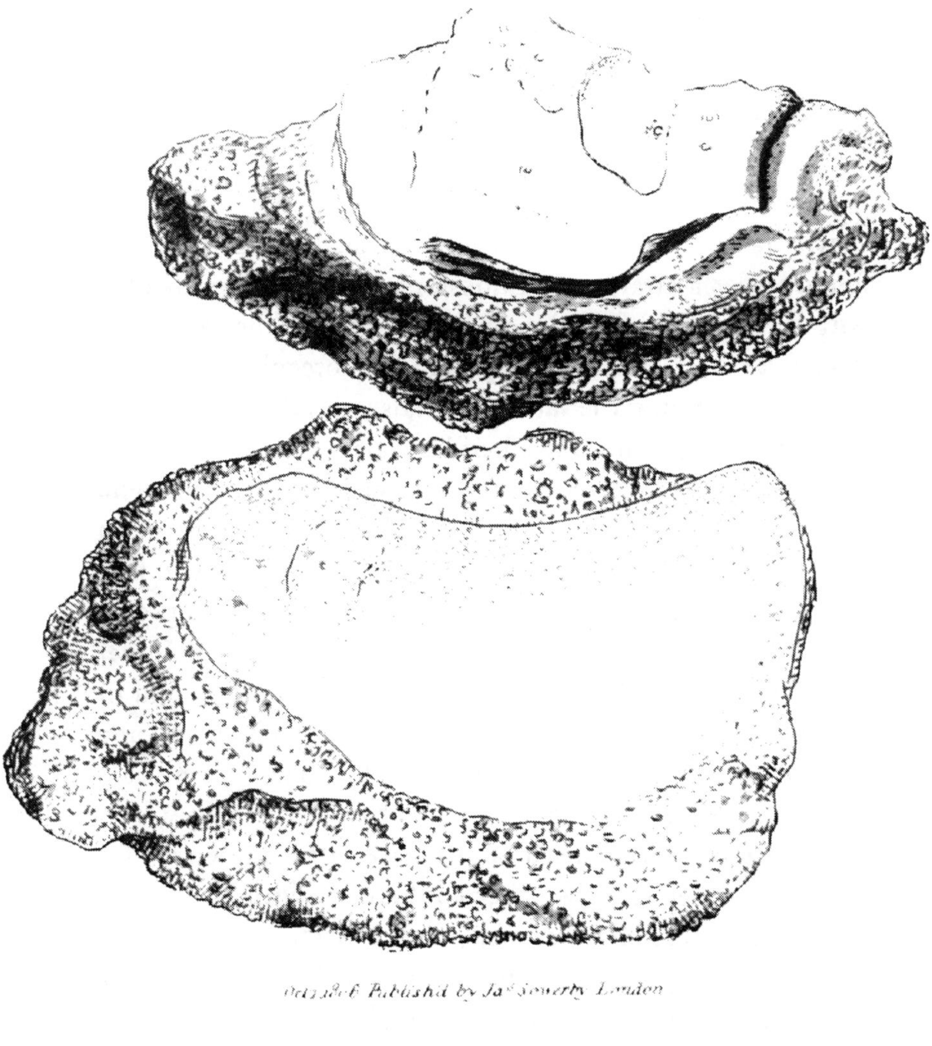

Oct.1.1816 Publishd by Jas Sowerby London

TAB. CLXXXV.

SILEX talcum, *var.* arenaceum.

Sandy Talc.

SYN. Mulatto Stone of the Irish.

THIS is abundant in the neighbourhood of Belfast, and found under a stratum of Limestone. It is harder than the last, being sufficiently incorporated and indurated to be termed a stone, thus differing from the foregoing, although perhaps containing the same materials; viz. Sand and Lime, with, possibly, some Clay, besides Talc.

The upper figure represents a piece sent by my friend Mr. Templeton, of Belfast, which has the impression of an oyster without any remains of the shell, and that part which seemed to be the impression of the connecting cartilage of the oyster has the fracture of Carbonate of Lime. There are often, I presume, very curious shells found in this substance; I, however, have, as yet, only obtained some crushed and unintelligible fragments. The green Talc which spots this stone gives it the characteristic by which it is commonly recognised, and I have often known stones with other greenish substances confounded with it.

The lower specimen was sent, with others, by Dr. Scott. All had more or less of impressions on them, and mostly of the same animal or shell, something like a curved oyster*,

* Of these I have some curious small specimens sent by my friend Mr. Wood, from near Wingham in Kent, about ten miles from the sea. There

which is very common where sand and chlorite are found together.

The formation of these rocks or sandy marles, &c. seems to be nearly of the same date wherever they are found, although sometimes near the surface of the earth, and sometimes covered with Limestone-rock at various depths.

TAB. CLXXXVI.

BITUMEN resiniferum.

Resinous Bitumen.

Class 1. Combustibles. *Order* 3. Mixed.
Gen. 1. Bitumen.

SYN. Retinasphaltum. *Hatchett in Phil. Trans. for* 1804. 410.

THIS very inflammable substance would, by its usual appearance, be taken for dark Umber while wet, and for common Clay when dry; consequently there is nothing in its common appearance that would indicate its inflammability or resinous quality. Very nice discrimination is therefore requisite to comprehend it. To the touch, however, it in some measure indicates a resinous quality.

Mr. Hatchett, who first mentioned it in *Linn. Trans.* *v.* 4. 139, observes that "a yellowish brown compact substance, which in colour and fracture somewhat resembles ferruginous clay, is found occasionally with the Bovey Coal. It is brittle, and is highly inflammable; it melts like Bitumen, and emits a smoke which in smell resembles Amber. This substance is but rarely found." He also observes, in the *Philosophical Transactions* for 1804, *p.* 402, that "it is found in pieces of a moderate size. The fracture is imperfectly conchoidal. It appears earthy externally; but when broken exhibits, in a slight degree, a vitreous lustre. The fragments are irregularly angular, and com-

does not apparently become softened when held some time in the hand, but emits a faint resinous odour. The specific gravity at a temperature of 60° of Fahrenheit is 1.135. When placed on a heated iron it immediately melts, smokes much, burns with a bright flame, and yields a very fragrant odour, like some of the sweet-scented resins, but which at last becomes slightly tainted with that of asphaltum. The melted mass, when cold, is black, very brittle, and breaks with a glossy fracture."

By the analysis of 100 grains by Mr. Hatchett it appears to contain:

Resin	55
Asphaltum	41
Earthy residuum	3
	99

I had the pleasure of receiving this, with a valuable series of the wood passing to the most perfect Bovey Coal, from Bovey Heathfield, near Chudleigh, Devonshire, by favour of Dr. Beeke.

TAB. CLXXXVII.

After figuring the Resinous Asphaltum, I conceive it very proper to show the nature of the Wood, from the same spot, in its passage to what the people of the neighbourhood call Plank, and Bovey Coal. Thus we have, by the favour of a good friend to science, obtained the Wood that may have assisted in affording this Resinous Asphaltum by a regular and gradual mineral change, depending upon

187

Oct. 1, 1806. Published by Ja.s Sowerby London

The upper specimen is nearly in the state in which we find very rotten wood sometimes above ground, and even in parts of living trees, with an earthy, fossil-like appearance. The grain and fracture of the wood still remain, with the fragments so sharp, that were it not for the colour and dull earthy appearance, it would seem but little altered: it is, however, so soft as to rub away under the finger like Roman Ochre, or the brown inside of the bark of some old firs. It burns at first with a flame, then with much smoke and an odour like the Resinous Bitumen, retaining a spark for some time, almost like Touchwood.

The lower specimen is nearly of the same nature as the above, with a more rotten appearance, and so soft as to have been pierced with roots like an earth; which also sometimes happens to the rotten parts of living trees, and I have seen their own branches shoot roots into such decayed parts. But what is very remarkable, the two broad surfaces of this specimen have exactly the appearance of having been burnt, so as to be a perfect charcoal; and neither these fibres nor the inner part seem to have been scorched*. Mr. Hatchett remarks that "the half-charred appearance of Bovey Coal cannot be adduced as any proof that the original vegetable bodies have been exposed to the partial effects of subterraneous fire." Now at first sight this specimen would seem to some an evidence to the contrary; but Nature coincides greatly with Mr. Hatchett, and in the general acceptation of our idea of water, there is little doubt but it has been, according to its nature, the cause of the effect here produced. We must remember that water is a very active agent, and is never quiet where there is the smallest room for

* Whether either of these specimens contains the alkaline principle or not I do not know. I am happy, however, to find that Mr. Hatchett, in his Analysis of the Iceland Schistus and Bovey Coal, *Phil. Trans.* for 1804, *p.* 399, found that the alkaline principle was wanting, as I asserted to be the case in the wood-like part of Newcastle Coal. *Brit. Min. tab.* 100.

its action; and as soon as the least fermentation is produced, calor or fire is evolved: and as this is continued from time to time, it is capable of producing the greatest effects. Thus the wood is formed into charcoal, slowly or otherwise, but actually by calor or fire, though water seems to be the principal agent. That this is one of Nature's methods for some of the varieties of combustion that take place will be perhaps more and more apparent, as well as that this agent, in many instances, produces common fire, as in a haystack, &c. yet a haystack may sometimes be in part so scorched and burnt as to be spoiled, without actual or rapid fire manifested by the red heat or flame.

Oct. 1. 1808. Publish'd by Jas. Sowerby London.

TAB. CLXXXVIII.

CARBO oxygenizatus bituminosus.
Bituminous Oxide of Carbon, or Bovey Coal.

Class 1. Combustibles. *Order* 3. Mixed.

SYN. Bovey Coal. *Hatchett in Phil. Trans. for* 1804. 385.

Compact Carbonated Wood. *Kirw.* 2. 61.

BOVEY, in Devonshire, has been some time famous for affording a fossilized wood of a nature rather peculiar to the place, commonly called Bovey Coal. I was prevented examining the place when I went that way, by want of time. Indeed there was little necessity for it; as the Rev. Mr. Rackett, Dr. Maton, Mr. Hatchett, &c. were more competent to the purpose: see *Maton's Tour,* 187. *Linn. Trans.* *v.* 4. 138. Although Mr. Hatchett examined the matter a second time, and would not venture to offer any opinion on the subject, I am highly indebted to my good friend the Rev. Dr. Beeke for procuring and assisting me with the best information concerning the particular nature of the change wrought on the wood as it advances towards the most perfect Bovey Coal.

It appears that the main dip is from south to north, and that, as Mr. Hatchett observes, the upper or southernmost end terminates like the remains of a bog, and the rest declines with an appearance of a continuation of this bog sunk downwards, forming the dip. *The upper part* contains stumps and broken remains of shrubs and trees, very little changed: *a little way down* they are somewhat resinously bituminized: see *tab.* 187. and the appearance of wood is still very distinct. To the touch they are greasy, and will readily polish with a slight application of the finger nail, being as it were saturated or closed with bituminous matter. In some places the resinous Asphaltum is intermixed; see *tab.* 186: in other places is seen the resino-bituminous wood, as it were burnt, or formed into common charcoal, *tab.* 187. The most perfect Bovey Coal is found at the greatest depth, being as it were a compound of these two, with the bituminous principle less resinous and the Coal more indurated, forming a wood-like bituminous coal. There are 17 strata found in intermediate gradations, composed of many varieties, depending on their situations: the lowest most perfect, about 70 feet deep or more, where the various pressures, and the state of confinement of the different strata, retard or accellerate the process. It seldom forms a large piece of black Surturbrand, although the charcoal before mentioned is as black as common charcoal. It, however, is often very dark-coloured, like part of the lowest figure, and is used as coal by the poor in the neighbourhood, and at a pottery, established on purpose, as I understand, to turn it to account. The smell, however, is offensive to most strangers, from its not being a true charcoal.

"It burns to a charcoal," as Mr. Hatchett observes, "readily, with a flame, like half-charred wood. It does not crackle; and, if completely burned, leaves a quantity of white ashes, exactly similar to those of wood."

If I might hazard a conjecture, the place seems suited to the phænomenon; the dip is from the south, where a wood and a bog (as I am informed) rise, as it were, out of the earth; or in other words, the northern side of this boggy wood is lowest, and has become gradually buried.

In the vicissitudes and changes that are natural to certain parts of the globe, we find plains formed between hills and mountains, lakes and rivers made level. Should this place formerly have been a large bog, the lowermost and deepest part of which, according to Dr. Maton, *v.* 1. 107, rests on a bed of sand now seventeen feet deep, it may have sunk into spaces for springs, &c. and that at seventeen different periods, between each of which there was time sufficient for the forming of fresh wood and fresh clay, in adequate proportion to the thickness of each stratum, as the sinking of the wood was probably followed by something of an inundation, whether quick or slow. Thus the change of the wood might be effected; the lowest or oldest being nearest, when compared with the rest, to perfect coal. The strata here are, however, less dense and looser than that under which the Northern or Newcastle Coals are formed; consequently it is a series pointing out something like a modern formation, giving us apparently a hint of the commencement of a more remote one. The cause, I think, may be presumed adequate to the effect; if we suppose a more or less perfect communication (according to the time of the falling of the first stratum) with the external southern aspect, the effects of wet and dry would rot the wood, and an evaporation would more or less be carried on, so as to accomplish the changes here seen; the upper strata being least changed, and only rotting by imbibing and evaporating moisture night and day, winter and summer. The resinous or bituminous parts have been more or less allowed to evaporate, according to the pressure; at the surface but little, and at the bottom scarcely at all, being out of the reach of perceptible change; and as it loses its most volatile principles, the more durable woody principle, or carbon, will last for ages, like a post charred to preserve it from the damps when put into the ground, which is so rendered extremely durable. We are sure to find that the post rots where there is most vicissitude of wet and dry, or nearly on the surface of the ground, and sometimes the damps will penetrate and rot the inside of the charred part which is itself almost indestructible.

TAB. CLXXXIX.

Surturbrand.

THIS is found near the mouth of the Ouse, ten miles from Brighton, Sussex, and is of the same kind as the Surturbrand of Iceland. Near the surface of the ground its changes are somewhat different from those of the foregoing, as it is less resino-bituminous, and more oxy-carbonized. It is found in large masses resembling compressed stumps of trees, and is of a most perfect black like Jet; but if compared with good Jet, has rather a gray cast. Sir Joseph Banks gave me a piece near two feet long, and above one in the broadest diameter; the compressed diameter is a little above two inches. The transverse section shows the concentric formation of the wood, and the sides crack and flake off more or less in circles. It is more brittle than Jet, and, being less bituminous, is not rendered electric by friction. It is often so much impregnated with Pyrites that it is apt to fall to pieces with the change of the atmosphere, even when preserved in cabinets.

Mr. Hatchett observes of Surturbrand, *Linn. Trans. v.* 4. 399, " that it is rather harder than Bovey Coal, but in every other respect the same." I take the liberty to say that it is generally blacker, and appears to have been less disturbed, being more regularly formed, and the transverse fracture more polished, while that of the Bovey Coal has rather a resinous appearance.

I have a specimen from Dr. Scott, found near Belfast, on the side of a mountain, the stratum being three feet thick, which comes immediately between the two (if we stand upon very nice distinction), not being so soft as Bovey Coal, nor having such a polish as the Surturbrand. Mr. Warburton was so good as to bring me a small series of

189

Oct. 1 1806 Publish'd by Jas. Sowerby London.

190

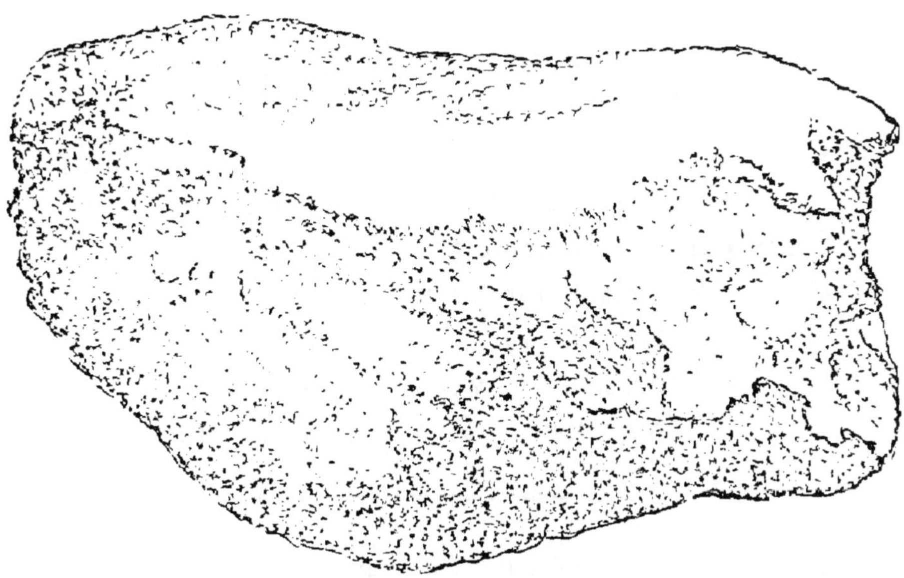

Oct.r 1 1806 Publish'd by Ja.s Sowerby London

TAB. CXC.

SULPHUR nativum.

Native Sulphur, or Brimstone:

Class 1. Combustibles.	*Order* 1. Homogeneous.
Gen. 6. Sulphur.	*Spec.* 1. Native.

GEN. CHAR. Solid. Colour pale yellow. Burns with a blue flame and pungent suffocating odour.

SPEC. CHAR. Uncombined.

SYN. Native Sulphur. *Kirw.* 2. 69.
Soufre. *Haüy*, 3. 277.
Natürlischer schwefel. *Emmerl.* 2. 89.
Pyrites nativus. *Linn. ed.* 13. *v.* 3. 113.

I HAVE specimens of native Sulphur from Amlwch, North Wales, by favour of the Rev. H. Davies. It is in an earthy-looking state, something like Flowers of Sulphur. I do not know that it has yet been found crystallized in Great Britain, but I however think there may still be a chance of it. The present is in a dull dusty state, but in some parts tolerably pure, and after being refined is cast into cones and sent up to London.

Upon a minute examination with a lens, it appears to be mixed with fine white Sand or granules of Quartz; see *the lower figure*.

It is rather curious to find that Sulphur with Iron, &c. is very common, and the odour of Sulphur is very strong in many places under ground fresh dug, particularly among decayed vegetables in summer; yet most authors speak of it as only found in the neighbourhood of volcanoes. This is not the case in Wales.

It is so well known in housewifery that it hardly requires identifying: I, however, add a part of Kirwan's description. Its colour is yellowish with some shade of green; it is found concrete or in loose powder; by friction emits a peculiar odour*, and becomes electric; heated to 170° Fahr. it generally evaporates; melts at 185°, and then appears red; it flames of a bright blue at 302°, emitting a sharp or pungent odour when it absorbs the pure air of the atmosphere, causing a stifling sensation, and becomes acidified, forming Sulphuric Acid; in close vessels it sublimes without much alteration.

* Pretty well known.

TAB. CXCI.

When I first visited the Isle of Dogs, at Blackwall, I thought it would be interesting to observe and collect the strata as belonging to a certain level pretty well determined by its vicinity to the Thames. Among other subjects it was both beautiful and interesting to observe the Sulphur forming on the old stumps of the trees that were found from nine to eighteen feet below the common level of the place; and perhaps it is no less remarkable that a stratum of leaves, which was in some places three or more feet thick, had a strong odour of Sulphur; but the Sulphur in general formed Sulphate of Lime, or Gypsum, with the little Lime among it, sparkling in the sun like minute diamonds. The same occurred sometimes upon the stumps of trees; but in some

291

Published by Jas Sowerby London.

parts the Sulphur was nearly pure, and even of a brighter colour than that from Amlwch, and covered largish spaces somewhat copiously, so that the wood was not discerned; in other places it seemed to be passing with the moisture out of the cracks, and followed the longitudinal direction of the fibres, very prettily contrasting with the sparkling Gypsum. The golden, if not shining, hue in some places equalled in beauty the yellow Lichens which often enrich the stumps of trees above ground. We believe, however, that it in some places occurs in the inside of growing trees, as we remember having seen it, but did not at the time make a careful investigation of the matter.

292

Oct 1 1806 Publish'd by Ja^s Sowerby, London

TAB. CXCII.

CARBO oxygenizatus.

Oxide of Carbon, or Coak.

Class 1. Combustibles.	*Order* 1. Homogeneous.
Gen. 7. Carbon.	*Spec.* 3. Oxide.

I HAVE been favoured by different friends with Coak, or what is commonly termed Cinder, found near the Dyke of Whinstone*, or Blue-stone Dyke, which crosses Cockfield Fell, and other Coal-mines in the North. When the Coal in these mines is examined, we find it is the more like Coak or Cinder the nearer it is to the Whinstone. The figure will show the Coak-like fracture on the Coal at *the upper part,* which is the appearance it had when I first received it, and exactly corresponded with many more in the general nature and appearance of artificial Coak, in forming a horizontal columnar appearance, with cracks, and burning without flame, &c. When examined with a lens, I found the perforations differing from artificial Coak, in being more smooth and shining. *The other part* of the Coal was not unlike common Coal at first appearance; but on nicer examination with a lens, I perceived little globules of Pyrites fitting to, and filling the hollows, and giving an idea that by their passing out the Coal would be rendered all like *the upper part.* I put it into a drawer with the other spe-

* Whin here must not be confounded with small-grained Granite: it is [illegible] basalt including feldtspar. This dyke is of a very great extent.

cimens in October 1804, and seldom looked at it till it became conspicuous this year, when the Pyrites had decomposed, and the Sulphur was passing out as figured, but with a different effect to what it seemed to have had in its native place; for, instead of leaving the Coal in the form of Coak, it was actually falling to pieces with another sort of fracture and tendency to crumble to dust, as it seems likely to do now very soon. *The lower figure* shows a fragment partly columnar, with the empty holes; *the left hand end* shows the Pyrites in the upper holes and the Sulphur below, corresponding with the figure of *the little upper* particle of iron Pyrites, and the Sulphur dust underneath. The Coal further from the Whin has fewer, but often larger, holes in it, and it sometimes has only a few. At the distance of a few feet the Coal is free from hollows, and burns like the best Newcastle Coal.

193

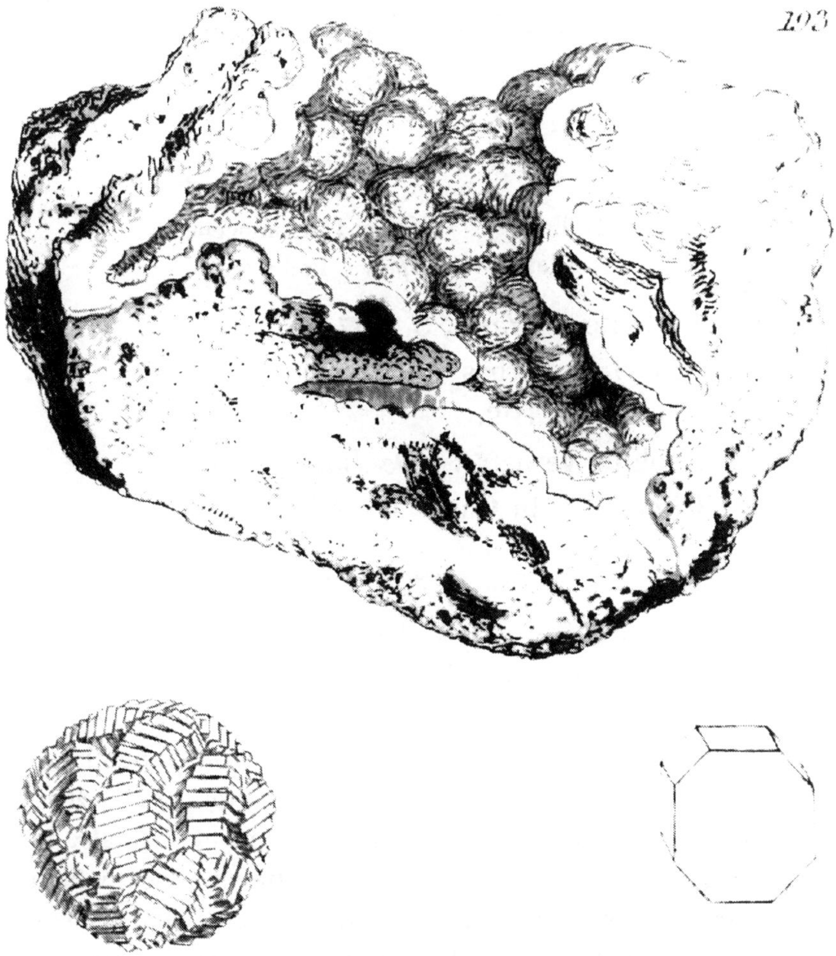

Dec.r [illegible] Published by Ja.s Sowerby, London

TAB. CXCIII.

SILEX Prehnites.

Prehnite.

Class 2. Earths. *Order* 1. Homogeneous.
Gen. 4. Silex. *Spec.* Prehnite.

SYN. Prehnite. *Haüy*, 3. 167. *Kirw.* 1. 274.
Schorl en gerbes de Schreiber, Prehnit. *Emmerl.* 1. 192.
Zeolithe verdatre. *De Born*, 1. 203.
Chrysolite du Cap. *De Lisle*, 2. 275.

PREHNITE was said to have been first discovered by colonel Prehn, who brought it from the Cape of Good Hope; Haüy, however, says that citizen Rochon was beforehand with him.—It is now found in many parts of Scotland. The present specimen comes from Dumbarton, and, when examined, exhibits a curious arrangement of crystals, in groups, forming hemisphærically, and showing faces which are the edges of tabular crystals arranged somewhat in segments of circles, having the larger crystals in the centre; see *the left-hand lower figure.* These crystals seem to betray signs of eight faces besides the two broader ones, like *the right-hand figures.* The nucleus we could not positively determine; it seems, however, to be a nearly rectangular table, as expressed by the dotted lines: thus the corners are as it were unfinished, or truncated. *The upper figure* is of the natural size; *the lower one,* for the sake of explanation, is magnified. This fossil lines the cavities of a sort of *Kragg* of *Kirwan*. The same crystals are also found at Salisbury Craig, Edinburgh. The substance is sometimes found amorphous, as at King's Park.

TAB. CXCIV.

THIS was sent me by my friend G. Laing, Esq., from Hartfield near Paisley. It is a fine specimen showing the green side of the radii, and the crystals in nearly regular four-sided columns, with two opposite truncations at the apex; these truncations, or secondary faces, are the same as those in the former description, *tab.* 193, on the upper edges, and ought to be particularly remembered, as they assist in forming a very singular modification, which will be shown in

TAB. CXCV.

THIS specimen came from the same place as the last, and leads to a curious modification. The crystals are arranged in double stellated groups, the radii of each extending so as to meet at the edges, like the spokes of two wheels placed against each other, contrary to their position on carriages; the periphery of the wheels coming together; the edges forming one circular face; see *the left-* and *right-hand figures.* These are grouped among roughish indistinct semi-orbicular masses. The colour is not so attracting as the formation, being dull and brownish. This mineral is formed in the cavities of Kragg rock somewhat approaching Porphyry. The crystals are nearly similar to those of *tab.* 193, but the secondary faces being larger, the terminal one is lost in an edge.

Dec 1 1806 Publishd by Jas Sowerby London

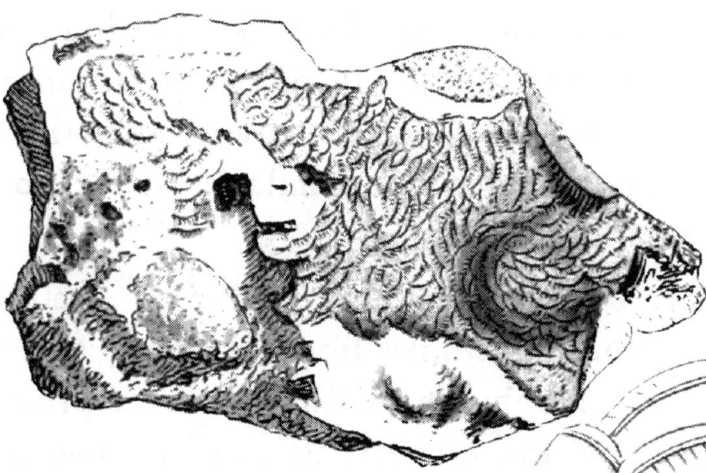

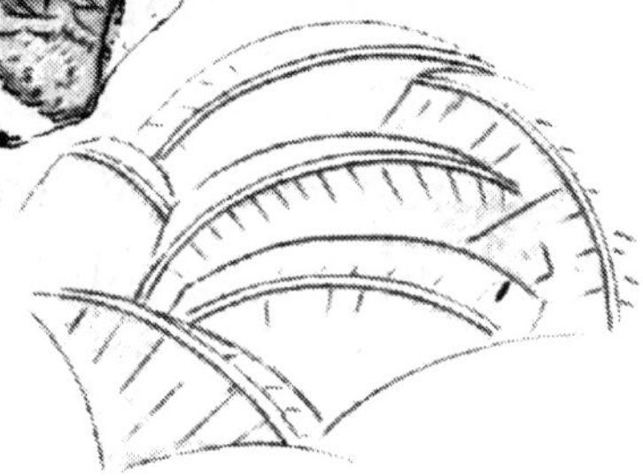

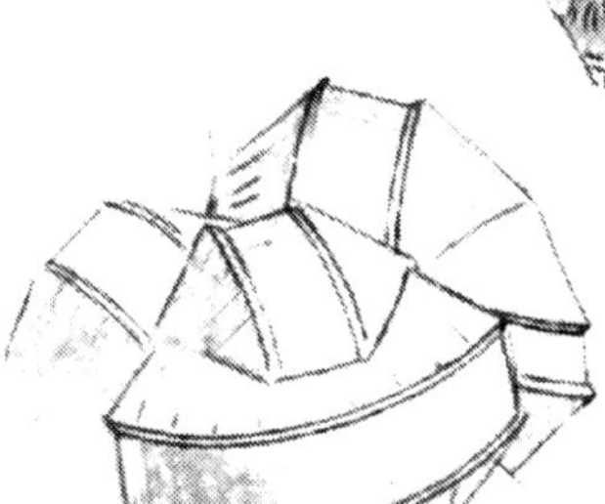

TAB. CXCVI.

THESE specimens are from Salisbury Craigs, Edinburgh, and show the conical wheel-like formation nearly in perfection, but very small. I have, therefore, greatly magnified *the right-hand outline*, which shows a narrow primitive edge, bounded by two lines. A similar crystallization is found at Kings Park, Edinburgh.

The lower sort I have from Frisky Hall, near the banks of the Clyde, about three miles from Glasgow; and also from Arthur's Seat, Edinburgh. It shows a broad primitive face on the edge of the crystals, bounded by two acute ridges, which are placed very confusedly on the specimens.

The Right Hon. Charles Greville has in his collection a specimen from Dauphiné with crystals, not much unlike the magnified *left-hand bottom figure*, in size and colour.

197

Dec. 1806. Published by Ja.s Sowerby, London.

TAB. CXCVII.

FERRUM suboxygenatum.

Suboxide of Iron. Magnetic Iron Ore.

Div. 3. Amorphous, in Grains.

A FEW years since, the Rev. Dr. Charles Sutton sent me, among other natural curiosities, some sand from Hunstanton, in Norfolk, of a blackish appearance. Not being at that time engaged in mineralogical pursuits, I laid it by; but when I was about to begin this work, he kindly brought me some more, which was still blacker. On looking attentively at it, I thought attractible iron might be the cause of its blackness, and soon found it to be the case, and that some of the sand might be rendered magnetical, the particles clinging together after being detached from the magnet. I considered it as a great curiosity, particularly as I had no where seen mention made of such sand, excepting from America, and requested to have more, at the same time speaking of its qualities; which request was kindly attended to. Shortly after, I received a note informing me that what had since been found was not attractible, and the same was observed by some other friends, who attempted to procure me some from other places. I examined the sand to discover the natural cause for this, and found it was not so black, and that there was little or no attractible iron; only some brown ochraceous particles, like Bog Iron Ore. It might have become oxidated by the weather: thus there is a season to find it in perfection, which should be attended to; and if it proves useful, it may, no doubt, be had in great abundance, being very common-looking sand sent me from Scarborough by my friend Dr. Travis, since which I have had the pleasure of receiving Iron Sand from Wicklow in Ireland, by favour of Dr. Scott. In examining this it evidently betrayed its habitat by minute particles of gold, besides which it has Octaëdral Iron more or less oxidated, and some Cubic Pyrites, Pebbles, &c.

Again, in July last, my valued friend James Brodie, Esq. sent me some beautiful *Arena ponderosa*, as it was called, sent by Governor Stewart of Fort St. George from the Ferry of Ardentenny in Argylshire, where, the governor observes, it is found in the greatest abundance, washed out of the banks by the sea; and what was sent me was as pure as that separated from the others, and somewhat brighter, with octaëdral crystals, small, black, abundant, and very attractible.

The right-hand upper figure is of the pebbles and sand from Hunstanton in a heap, and the particles of iron of the size generally found, by the side. *The figure underneath* is such as is found when the sea has washed it and oxidated it. *The five pebbles on the side magnified* are three common Quartz pebbles, which compose common sand, and two darker, chiefly Oxide of Iron. *The left-hand upper figure* is the Irish Sand, and *the figures on the side* show the Octaëdron, Cube, Gold, and Pyrites. Beneath is the outline of a magnet, and the Iron in common as attracted by it at the base. *The middle figure* is a parcel like that from Scotland, or such as has been separated from the other sands.

Thus it appears that this sort of Sand, which was hitherto supposed to be found only in America, has now been found in England, Scotland, and Ireland, or the three grand divisions of the British empire. I should think these might be of great use to the proprietors, if properly attended to. I understand Ardentenny is the property of Lord

198

TAB. CXCVIII.

CALX carbonata, *var.* ferrifera.
Ferriferous Carbonate of Lime.

THIS is part of a superb specimen collected last summer by John George Children, Esq. at Audlim mine, about eight miles from Bodmin, Cornwall, and is somewhat the more remarkable as it used to be said that Carbonate of Lime was not to be found in Cornwall. The whiter part of the specimen is a cavity handsomely filled with crystallized Carbonate of Lime of a very uncommon modification, being nearly an hexaëdral plate with the equiaxe and primitive bevellings, if I may so call them. It is rather remarkable that the external surfaces of these crystals are whitish, and the inside of a rich dark brown, as the darker surrounding part shows*. *Tab.* 62. British Mineralogy is nearly of the same nature, but under common circumstances grows blacker by exposure to the air.

This specimen has many other curious circumstances of change and position of mineral substances attending it; viz. the redder parts are a sort of Carnelian Quartz, somewhat approaching Chalcedony, coloured by a rich Oxide of Iron, and this is sometimes covered by Cachalon: see British Mineralogy, *tab.* 111.

* This specimen has probably been broken from an opposite piece, and

Besides this there are yellow spiculated tubes, almost crystallized, radiating, &c. These are to be seen as forming over wire-shaped Pyrites; see *Brit. Min. tab.* 162. This has decomposed in some parts, leaving the hollow where it has been with enough to show the appearance of a wire, as the fracture in some parts on the opposite side shows; perhaps it may be between *Eisen Keisel,* or the German Iron Flint, and Carnelian. Some gray Cachalon covers the Carnelian in the hollow as represented at the top of the figure. The yellower Quartz seems to be coloured by yellow oxide of Iron, probably the decomposed Pyrites.

TAB. CXCIX.

SILEX Quartzum.

Crystallized Quartz.

Div. 1. Crystallized.

THESE curious specimens were lent me by Philip Rashleigh, Esq. of Menabilly, and are said to contain Arsenical Cobalt. They came from Poulteney mine, and are rare specimens, on account of the mine having been destroyed by the overflowing of a river. The Quartz is more regularly eighteen-sided than usual, although it is as it were clogged up with such abundance of a metal, that it otherwise would be nearly obliterated in its character; and that it should crystallize under such circumstances, distinctly and regularly at both ends, merely as if heaped on each other, as in *the top figure*, is very remarkable. There appears to be little else than Arsenical Iron among them, but capillary Silver and flowers of Cobalt are sometimes to be seen very distinctly about the gangue.

TAB. CC.

FERRUM sulphureum.

Iron Pyrites in petrified Wood.

THE upper specimen seems to have been part of a cylindrical piece of wood, and was found 260 feet below the surface of the earth in digging a well in Richmond Park in 1804. It appears to have had worm-holes, or holes of *Terebellæ*, perforating it in various directions, which may be presumed to have happened before the process called petrifying had taken place. This may more properly be called Pyritaceous Wood, as the Pyrites or Sulphuret of Iron has filled the pores of the wood so perfectly, that the shape, and somewhat of the texture of wood, was seen, but as if formed of Pyrites. The worm-holes, some lined with Pyrites, and others doubly lined. One side being nearly covered with Pyrites makes it a beautiful specimen, as well as an instructive one. *The lower piece* was perhaps of different wood, appearing like part of a plank. This was found 100 feet deep in digging a well for Mr. Truman's brewhouse, Spitalfields, London. The worm-holes are lined like the other; but they seem to have been a particular species which prefer a straight direction, crossing the fibres of the wood. It is somewhat remarkable that the woody nature remains, and being found damp and fully saturated with the Pyrites, it cracked and contracted from it, and is held together in some parts as if artificially done with wires, and is in some parts curved and warped. Thus, although these pieces of wood seem to have been many years under this process, they are not much changed, but I may say rather, preserved, as, now it is exposed to the common air, these changes which show its nature have become evident; it will soon fall to decay, in the same manner as some part has already done. The Iron and Sulphur decomposing the water of the atmosphere, the Sulphur becoming acidified dissolves the Iron, forming green vitriol or Sulphate of Iron which is very deliquescent.

SYSTEMATICAL INDEX

TO

VOL. II.

ALPHABETICAL INDEX

TO

VOL. II.

E

F

R. Taylor and Co., 38, Shoe Lane.

CORRIGENDA.

Page.	line.	
2*	18	*for* Silica, *read* Silex.
	24	*for* Silica, *read* Silex.
3	21	*for* Siberlan, *read* Siberian.
9	2	*for* Ferrum argillaceum, *read* Ferrum oxygenatum; *var.* argillaceum.
	5	*for* Spec. 1. Argillaceous, *read* Spec. 3. Oxide of
11	2	*for* F. L. S. *read* Tr. L. S.
15		*dele from,* "They are," *on l.* 12 *as far as* "spiculæ," *on l.* 14. *and read* "They form an irregular echinated sphere, on a gangue of Galæna or Sulphuret of Lead." *And on l.* 15 *add* "They are found to be fusible by the blowpipe."
31	5	*for Gen.* 6. read *Gen.* 5.
57	3	from the bottom *dele* "and 131."
	13	*for* Gorm *read* Gorum.
87	1	*for* 147 *read* 146.
95	3	*for* Ironey *read* Irony.
101	5	*for Spec.* 4. *read Spec.* 5.
131	5	*from bottom for* species, *read* variety.
135	2	*for* sulphuretum, *read* sulphureum.
137	3	*for* stellate, *read* stellated.

Lightning Source UK Ltd.
Milton Keynes UK
UKHW050930251022
411061UK00009B/744

9 781241 497088